Introductory
Discrete Mathematics

Introductory
Discrete Mathematics

V. K. Balakrishnan

DOVER PUBLICATIONS, INC.
New York

Copyright

Bibliographical Note

This Dover edition, first published in 1996, is an unabridged, corrected republication of the work first published by Prentice Hall, Englewood Cliffs, N.J., 1991.

Library of Congress Cataloging-in-Publication Data

Balakrishnan, V. K., date.
    Introductory discrete mathematics / V. K. Balakrishnan.
        p.   cm.
    "An unabridged, corrected republication of the work first published by Prentice Hall, Englewood Cliffs, N.J., 1991"—T.p. verso.
    Includes bibliographical references (p.   –   ) and index.
    ISBN-13: 978-0-486-69115-2 (pbk.)
    ISBN-10: 0-486-69115-2 (pbk.)
    1. Mathematics.   2. Computer science—Mathematics.   I. Title.
QA39.2.B357   1996
511—dc20   95-52384
                                                                      CIP

Manufactured in the United States by LSC Communications
69115215   2020
www.doverpublications.com

To Geeta

Contents

1   Combinatorics   35

2   Generating Functions   80

3   Recurrence Relations   94

4   **Graphs and Digraphs   120**

5   **More on Graphs and Digraphs   140**

6   **Trees and Their Applications   164**

Preface

Introductory Discrete Mathematics is a concise text for a discrete mathematics course at an introductory level for undergraduate students in computer science and mathematics. The essential components of any beginning level discrete mathematics curriculum are combinatorics, graph theory with applications to some standard network optimization problems, and algorithms to solve these problems. In this book the stress is on these core components. Both the Association for Computing Machinery and the Committee for the Undergraduate Program in Mathematics recognize the vital role of an undergraduate course in discrete methods that introduces the student to combinatorial mathematics and to algebraic and logical structures focusing on the interplay between computer science and mathematics.

The material in Chapter 0 serves as an introduction to the fundamental operations involving sets and the principle of mathematical induction. For those students familiar with the topics discussed here, this is essentially a chapter for review.

The standard topics in combinatorics in any course on discrete mathematics are covered in Chapters 1, 2, and 3. These topics include basic counting principles, permutations, combinations, the inclusion–exclusion principle, generating functions, recurrence relations, and an introduction to the analysis of algorithms. The role of applications is emphasized wherever possible. There are more than 200 exercises at the end of these chapters. Each counting problem requires its own special insight, and it is advantageous for the student to work out several of these problems.

In the next three chapters is a survey of graphs and digraphs. We begin with treating graphs and digraphs as models of real-world phenomena by giving several examples. The connectedness properties of graphs and digraphs are studied. Basic results and applications of graph coloring and of Eulerian and Hamiltonian graphs are presented with a stress on applications to coding and other related problems. Two important problems in network optimization are the minimal spanning tree problem and the shortest distance problem; they are covered in the last two chapters. The approach to compute the complexity of algorithms in these chapters is more or less informal.

A very brief nontechnical exposition of the theory of computational complexity and NP-completeness is outlined in the appendix.

It is possible to cover the topics presented in this book as a one-semester course by skipping some sections if necessary. Of course it is for the instructor to decide which sections she or he may skip.

My chief acknowledgment is to the students who have studied discrete mathematics with me at the University of Maine during the past decade. They taught me how to teach. Their contributions and encouragement are implicit on every page. In particular, I would like to mention the names of Rajesh and Thananchayan. My scientific indebtedness in this project encompasses many sources including the articles and books listed in the bibliography. If there are errors or misleading results, the blame of course falls entirely on my shoulders. Finally, it goes without saying that I owe a great deal to the interest and encouragement my family has shown at every stage of this work.

V. K. Balakrishnan

CHAPTER

0

Set Theory
and Logic

0.1 INTRODUCTION TO SET THEORY

The concept of a set plays a very significant role in all branches of modern mathematics. In recent years set theory has become an important area of investigation because of the way in which it permeates so much of contemporary mathematical thought. A genuine understanding of any branch of modern mathematics requires a knowledge of the theory of sets for it is the common foundation of the diverse areas of mathematics. Sets are used to group distinct objects together. It is necessary that the objects which belong to a set are *well-defined* in the sense that there should be no ambiguity in deciding whether a particular object belongs to a set or not. Thus, given an object, either it belongs to a given set or it does not belong to it. For example, the first five letters of the English alphabet constitute a set which may be represented symbolically as the set {a, b, c, d, e}. An arbitrary object belongs to this set if and only if it is one of these five letters. These five distinct objects can appear in any order in this representation. In other words, this set can also be represented by {d, b, a, e, c}. The objects that belong to a set need not possess a common property. Thus the number 4, the letter x, and the word "book" can constitute a set S which may be represented as $S = \{x, \text{book}, 4\}$. A particular day may be cold for one person and not cold for another, so the "collection of cold days in a month" is not a clearly defined set. Similarly, "the collection of large numbers" and "the collection of tall men" are also not sets.

The term *object* has been used here without specifying exactly what an object is. From a mathematical point of view, *set* is a technical term that takes its meaning from the properties we assume that sets possess. This informal description of a set, based on the intuitive notion of an object, was first given by the German mathematician Georg Cantor (1845–1918) toward the end of the nineteenth century and the theory of sets based on his version is known as *naive set theory.* In Cantor's own words, "a set is bringing together into a whole of definite well-defined objects of our perception and these objects are the elements of the set." The sets considered in this book can all be viewed in this framework of Cantor's theory.

Thus a **set** is a collection of distinct objects. The objects in a set are called the **elements** or **members** of the set. If x is an element of a set A, we say that x **belongs** to A, and this is expressed symbolically as $x \in A$. The notation $y \notin A$ denotes that y is not an element of the set A.

Finite and Infinite Sets

A set is **finite** if the number of elements in it is finite. Otherwise, it is an **infinite** set. The set of positive integers less than 100 is a finite set, whereas the set of all positive integers is an infinite set. If X is a finite set, the **cardinality** of X is the number of elements that belong to X, and this nonnegative integer is denoted by $N(X)$. A set of cardinality 1 is called a **singleton set.**

If a set is finite and if its cardinality is not too large, we can describe it by enumerating all its elements. For example, the representation $S = \{a, e, i, o, u\}$ describes the set of all vowels of the English alphabet. If the cardinality is too large, this enumerative method is not very convenient. In some cases, if there is no ambiguity we make this enumerative description more concise. For example, the set D of positive integers between 25 and 123 can be represented as $D = \{25, 26, 27, \ldots, 121, 122, 123\}$. A better way of representing D is by stating the property for its membership. An object n is an element of this set D if and only if n is a positive integer that is at least 25 and at most 123. In other words, we write

$$D = \{n : n \text{ is a positive integer}, 24 < n < 124\}$$

The infinite set N of all natural numbers can be represented unambiguously as $N = \{1, 2, 3, \ldots\}$ or as $N = \{n : n$ is a natural number$\}$ by stating its membership criterion. The notation of representing a set by stating the criteria of its membership as described above is called the **set-builder notation.**

Subsets of a Set and the Empty Set

A set P is a **subset** of a set Q if every element of P is an element of Q. We use the notation $P \subset Q$ to denote that P is a subset of Q. A subset of a subset is no doubt a subset. When P is a subset of Q, we may say that Q **contains** P and that P **is contained in** Q. By our definition every set is a subset of itself. The set P is a **proper subset** of Q if (1) P is a subset of Q and (2) there is at least one element of Q that is not an element of P. The set of positive integers is a proper subset of the set of all real numbers.

If A is a subset of B, the **relative complement of A in B** is the set of elements in B that are not elements of A. The relative complement of A in B is denoted by $B - A$ and it can be described by its membership criterion as

$$B - A = \{t : t \in B, t \notin A\}$$

Two sets are **disjoint** if they have no elements in common. On the other hand, two sets are **equal** if they have the same elements. We write $X = Y$ when the sets X and Y are equal. Obviously, two sets are equal if and only if each is a subset of the other. For instance, if $X = \{r : r$ is a root of the $x^2 - 5x + 6 = 0\}$ and $Y = \{2, 3\}$, then $X = Y$.

A set is **empty** if it has no elements. A fact emerges that some people find surprising: there is only one empty set. (Suppose that E and F are two empty sets. If they are not the same, they are not equal. So one of them should have at least one element that does not belong to the other. So one of the two sets is not empty. This contradicts the assumption that both E and F are empty.) The unique **empty set** (or **null set**) is denoted by ϕ. The fact that the empty set is a subset of any set is established by "vacuous reasoning": If it were not a subset of a given set S, there should be at least one element in the empty set that is not in S. In particular, there should be at least one element in the empty set that is a contradiction. Of course, a set is empty if and only if its cardinality is zero.

In some cases we will be considering sets that are all subsets of a set U which is called the **universal set.** For example, if the sets under consideration are A, B, and C, where $A = \{3, 8, 6, 7, x\}$, $B = \{8, 4, y, t, 5\}$, and $C = \{3, 4, x, t, 9\}$, then any set containing the set $D = \{3, 8, 6, 7, x, 4, y, t, 5, 9\}$ can be considered as a universal set as far as A, B, C, and D are concerned.

Once the universal set U is fixed, the relative complement of a subset A in U is called the **absolute complement of** A and is denoted by A^c. Thus if the universe is the set of all nonnegative integers and E is the set of all even numbers, then E^c is the set of all odd numbers. Observe that the absolute complement of the absolute complement of any set A is the set A itself.

The Power Set of a Set

A set can have other sets as its elements. For instance, the set S consisting of the letter x, the set $\{a, b\}$ and the number 4 is represented as $S = \{x, \{a, b\}, 4\}$. A set of subsets is also known as a **class** or **family** of sets. The class of all subsets of a given set X is called the **power set** of X and is denoted by $P(X)$. For example, if $X = \{1, 2\}$, the elements of $P(X)$ are the empty set, the singleton set $\{1\}$, the singleton set $\{2\}$, and the set X. Thus $P(X) = \{\phi, \{1\}, \{2\}, \{1, 2\}\}$.

Cartesian Products of Sets

The **ordered** n**-tuple** $(a_1, a_2, a_3, \ldots, a_n)$ is a collection of the n objects a_1, $a_2, \ldots, a_n$ in which a_1 is the first element, a_2 is the second element, $\ldots$, and a_n is the nth element. In an ordered n-tuple, the elements need not be distinct. A set with n elements is thus an *unordered* n-tuple of n distinct elements, since in a set the order in which the elements are considered is irrelevant. An ordered 2-tuple is called an **ordered pair.** Two ordered n-tuples $(a_1, a_2, \ldots, a_n)$ and $(b_1, b_2, \ldots, b_n)$ are said to be equal if $a_i = b_i$ for $i = 1, 2, \ldots, n$. The set of all ordered pairs (a, b), where a is an element of a set A and b is an element of a set B, is called the **cartesian product** of A and B and is denoted by $A \times B$. In other words,

$$A \times B = \{(a, b) : a \in A \text{ and } b \in B\}$$

For example, if $A = \{1, 2\}$ and $B = \{1, 3\}$, then the cartesian product $A \times B$ is the set $\{(1, 1), (1, 3), (2, 1), (2, 3)\}$. More generally, the cartesian product of the sets $A_1, A_2, \ldots, A_n$ denoted by $A_1 \times A_2 \times \cdots \times A_n$ is the set of all ordered n-tuples of the form $(a_1, a_2, \ldots, a_n)$, where a_i is any element of A_i $(i = 1, 2, \ldots, n)$.

Intersections and Unions of Sets

There are two important constructions that can be applied to subsets of a set to yield new subsets. Suppose that A and B are two subsets of a set X. The set consisting of all elements common to both A and B is called the **intersection** of

A and B and is denoted by $A \cap B$. Obviously, the intersection of a set and the empty set is the empty set and the intersection of any set A and A is A. Also, the intersection of a set and its absolute complement is empty since no element can be simultaneously in A and not in A. Moreover, it follows from the definition that set intersection has the commutative property: The intersection of A and B is equal to the intersection of B and A.

The set consisting of all elements that belong to either A or to B or to both A and B is called the **union** of A and B and is denoted by $A \cup B$. The union of a set A and the empty set is the set A and the union of A and A is also A. Set union also is commutative: $A \cup B = B \cup A$. More generally, the intersection of a class of sets is the set of elements (if any) that belong to every set of the class. The union of a class of sets is the set of those elements that belong to at least one set in the class. It is an immediate consequence of the definition that both set intersection and set union possess the associative property: (1) $A \cap (B \cap C) = (A \cap B) \cap C$ and (2) $A \cup (B \cup C) = (A \cup B) \cup C$. So the former can be written as $A \cap B \cap C$ and the latter as $A \cup B \cup C$ unambiguously.

Two sets are disjoint if and only if their intersection is empty. A class of sets is **pairwise disjoint** if the intersection of any two sets in the class is empty. A class $C(X)$ of subsets of a set X is called a **partition** of X if (1) $C(X)$ is pairwise disjoint, and (2) the union of the sets in $C(X)$ is the set X. For instance, the class $\{\{2, 4\}, (1, 3, 5\}, \{6\}\}$ is a partition of the set $\{1, 2, 3, 4, 5, 6\}$.

Venn Diagrams of Sets

A very useful and simple device to represent sets graphically for illustrating relationship between them is the **Venn diagram,** named after the English logician John Venn (1834–1923). In a Venn diagram, the universal set U that contains all the objects under consideration is usually represented by a rectangle, and inside this rectangle subsets of the universal set are represented by circles, rectangles, or some other geometrical figures. In the Venn diagram shown in Figure 0.1.1, we have three sets A, B, and C which are subsets of the universal set U. The drawing of the ellipse that represents the set A inside the ellipse that represents the set B indicates that A is a subset of B. The fact that A and C are

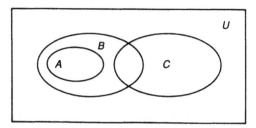

FIGURE 0.1.1

disjoint is made clear by representing them by two nonintersecting ellipses. The fact that the intersection of B and C is nonempty is made obvious by showing that the two ellipses which represent these two sets overlap each other. The region in the rectangle (which represents the universal set) that is outside the ellipses that represent the three sets is the absolute complement of the union of these three sets.

Distributive Laws and De Morgan's Laws

We conclude this section on sets with the following two theorems related to set operations involving intersections, unions, and taking absolute complements. These theorems can easily be established by drawing Venn diagrams. However, it is instructive to prove them without the aid of Venn diagrams, for in many cases it may not be possible to represent the sets under consideration by such diagrams, as we will see later in the book.

THEOREM 0.1.1　(Distributive Laws)

(a) $A \cap (B \cup C) = (A \cap B) \cup (A \cap C)$.
(b) $A \cup (B \cap C) = (A \cup B) \cap (A \cup C)$.

Proof:

(a) One way of showing that two sets are equal is by establishing that each is contained in the other. Let t be an element of $A \cap (B \cup C)$. Then t is an element of A and t is either an element of B or an element of C. In either case, t is an element of $A \cap B$ or $A \cap C$. In other words, $A \cap (B \cup C)$ is a subset of $(A \cap B) \cup (A \cap C)$. Next, suppose that t is an element of $(A \cap B) \cup (A \cap C)$. This implies that t is either in $A \cap B$ or in $A \cap C$. So t is necessarily in A and it is in at least one of the two sets B or C. Thus t is in A and also in either B or in C. In other words, t belongs to the intersection of A and to $B \cup C$. Thus $(A \cap B) \cup (A \cap C)$ is a subset of $A \cap (B \cup C)$.

(b) This is left as an exercise.　　　　　　　　　　　　　　　　　　　$\diamondsuit$

THEOREM 0.1.2　(De Morgan's Laws)

(a) $(A \cap B)^c = A^c \cup B^c$.
(b) $(A \cup B)^c = A^c \cap B^c$.

Proof:

(a) Let t be an element of $(A \cup B)^c$. Then t belongs to neither A nor B. So t is necessarily in both A^c and B^c. Thus $(A \cup B)^c$ is a subset of the intersection of A^c and B^c. On the other hand, if t is in the intersection of A^c and B^c, it is neither in A nor in B. This implies that t is not in the union of A and B. Hence the intersection of A^c and B^c is contained in the complement of $A \cup B$.

(b) This is left as an exercise.

0.2 FUNCTIONS AND RELATIONS

In this section a brief review of the basic ideas involving functions and relations is presented. The concept of a function is pivotal in mathematics.

The Domain and the Range of a Function

Let X and Y be two nonempty sets. A **function** f **from** X **into** Y, denoted by $f: X \rightarrow Y$, is a rule that assigns to every element in X a *unique* element in Y. The set X is the **domain** of the function and the set Y is its **codomain.** If y is the unique element in Y assigned by the function f to the element x, we say that y is the **image** of x and x is a **preimage** of y and we write $y = f(x)$. The set $f(A)$ of all images of the elements of a subset A of X is called the **image of the set** A. The set $f(X)$ is called the **range** of the function. The range of a function is a subset of its codomain. If y is an element in the range of f, the set of all the preimages of y is denoted by $f^{-1}(y)$. If A is a subset of the range $f(X)$, the **inverse image** of the set A is the set $\{x : x$ is in X and $f(x)$ is in $A\}$, which is denoted by $f^{-1}(A)$. If f is a function from X to Y, it is customary to say that f **maps** the set X into Y.

Example 0.2.1

Let R be the set of all real numbers.

(a) Let $f: R \rightarrow R$ be the function that assigns the real number $x + 1$ to each real number x. In other words, $f(x) = x + 1$. Here the domain, codomain, and range of f is R.

(b) Let $f: R \rightarrow R$ be the function defined by $f(x) = x^2$. So every real number is assigned to its square. Here the domain and codomain of f are R and its range is the set of all nonnegative numbers.

Example 0.2.2

Let $A = \{a, b, c\}$ and $B = \{1, 2, 3, 4\}$. Then the rule f defined by $f(a) = 1$, $f(b) = 1$, $f(c) = 4$, and $f(d) = 2$ is a function f from A to B. The range of f is $\{1, 2, 4\}$, which is a proper subset of its codomain B.

Surjections, Injections, and Bijections

A function $f: X \rightarrow Y$ is called a **surjection** if $f(X) = Y$ and we say that f is function from X **onto** Y. A function $f: X \rightarrow Y$ is called an **injection** (or a **one-to-one mapping**) if two different elements in X have two different images in Y. A function $f: X \rightarrow Y$ is a **bijection** if it is both a surjection and an injection. The bijection from a set X onto itself that maps every element in the set into itself is called the **identity mapping** i_x **on** X. Two sets are **equivalent** if there is a bijection from one to the other. It is evident that two *finite* sets are equivalent if and only if they both have the same cardinality.

Example 0.2.3

(a) Let $X = \{a, b, c\}$, $Y = \{p, q\}$, and $f: X \rightarrow Y$, where $f(a) = p$, $f(b) = q$, and $f(c) = p$. Then f is a surjection and f maps X onto Y. Here f is not an injection.

(b) If $X = \{a, b, c\}$, $Y = \{p, q, r, s\}$ and if $g(a) = p$, $g(b) = q$, $g(c) = r$, then g is an injection but not a surjection. The range $g(X) = \{p, q, r\}$ is a proper subset of the codomain Y.

(c) If $X = \{a, b, c\}$, $Y = \{p, q, r\}$ and if $h(a) = p$, $h(b) = q$, and $h(c) = r$, then h is a bijection.

(d) If R is the set of real numbers and $f: R \rightarrow R$ the function defined by $f(x) = x^2$, then f is neither a surjection, because no negative number has a preimage, nor an injection, because the image of x and the image of $-x$ are both equal.

The Inverse of a Function

Let $f: X \rightarrow Y$ be a bijection. The **inverse function** of f is the bijection f^{-1}: $Y \rightarrow X$ defined as follows: For each y in Y, we find that unique element x in X such that $f(x) = y$. Then we define $x = f^{-1}(y)$. A function $f: X \rightarrow Y$ is said to be **invertible** whenever its inverse exists.

Example 0.2.4

If $X = \{1, 2\}$, $Y = \{p, q\}$, $f(1) = p$, and $f(2) = q$, then f is a bijection from X onto Y and its inverse f^{-1} is the bijection from Y onto X that maps p into 1 and q into 2.

A function f whose domain X and codomain Y are subsets of the set R of real numbers is **strictly increasing** if $f(x) < f(y)$ whenever $x < y$ and **strictly decreasing** if $f(x) > f(y)$ whenever $x < y$. It follows from the definition that both strictly increasing functions and strictly decreasing functions are injections.

Compositions of Functions

Let $g: X \to Y$ and $f: Y \to Z$. The **composition** of f and g, defined by $f \circ g$, is a function from X to Z defined by $(f \circ g)(x) = f(g(x))$. In other words, the function $f \circ g$ assigns to an element x in X that unique element assigned by f to $g(x)$.

Example 0.2.5

(a) Let $X = \{a, b, c\}$, $Y = \{p, q, r, s\}$, and $Z = \{1, 2, 3\}$. Let $g(a) = p$, $g(b) = q$, and $g(c) = r$, so that $g(X) = \{p, q, r\}$. Then if $f: g(X) \to Z$ is defined by $f(p) = 1$, $f(q) = 2$, and $f(r) = 3$, we have

$$(f \circ g)(a) = f(g(a)) = f(p) = 1$$
$$(f \circ g)(b) = 2$$
$$(f \circ g)(c) = 3$$

(b) Let f and g be functions from the set of integers to the set of integers. If $f(x) = 4x + 3$ and $g(x) = 2x + 5$, then

$$(f \circ g)(x) = f(g(x)) = f(2x + 5) = 4(2x + 5) + 3 = 8x + 23$$
$$(g \circ f)(x) = g(f(x)) = g(4x + 3) = 2(4x + 3) + 5 = 8x + 11$$

If f is a bijection from X onto Y, its inverse is a bijection from Y to X. If $y = f(x)$, then $f^{-1}(y) = x$. Thus $f^{-1}(f(x)) = f^{-1}(y) = x$ and $f(f^{-1}(y)) = f(x) = y$. In other words, the composition of a bijection from X onto Y and its inverse is the identity mapping from Y onto itself.

Sequences, Strings, and Languages

A **sequence** is a function whose domain is a set of *consecutive* integers. If the domain X is a finite set of n integers, we may take $X = \{1, 2, 3, \ldots, n\}$ or $\{0, 1, 2, \ldots, n - 1\}$. Otherwise, we may take X as the set of natural numbers or as the set of nonnegative integers. If $f: X \to Y$ is a sequence, the image $f(i)$ of the integer i is sometimes written as f_i and is called the i**th term of the sequence.**

Notice that in representing a sequence s, the *order* in which the images under s appear is important. This is not so in the case of a function. For example, if f is the function from $X = \{1, 2, 3\}$ to $Y = \{p, q\}$, where $f(1) = f(2) = p$ and $f(3) = q$, the collection of the images of the three elements of X under f can be represented as p, p, q in any order. But the *sequence* f is represented as

$(f(1)f(2)f(3))$ or as (ppq). A sequence whose domain is finite consisting of n consecutive integers and whose codomain is Y defines a **string of length n in Y** or **word of length n in Y**. In fact, any such sequence is an n-tuple.

Example 0.2.6

(a) Let $X = \{1, 2, 3, \ldots\}$ and R the set of real numbers. Consider the sequence $f: X \rightarrow R$ defined by $f(n) = 1/n$. Then the nth term of the sequence denoted by f_n is the image $f(n)$ of the element n in X. This sequence is also denoted by $\{1/n : n = 1, 2, 3, \ldots\}$.

(b) Let $X = \{1, 2, 3, 4, 5\}$ and $Y = \{a, b, c, d\}$ and consider the sequence $f: X \rightarrow Y$ defined by $f(1) = a, f(2) = b, f(3) = a, f(4) = c$, and $f(5) = b$. Then this sequence is the string $abacb$ of length 5 in Y which is also the 5-tuple $(abacb)$.

Any mapping f from $A \times A$ into A is called a **binary operator on A**. For instance, if R is the set of real numbers, the mapping $f: R \times R \rightarrow R$ defined by $f(a, b) = a + b$ (which is, in fact, the addition operator) is an example of a binary operator on R.

If S is any nonempty set, we denote by S_n the set of all strings of length n in S and S^* the set of all strings (including the null string with no elements). Any subset of S^* is called a **language over the alphabet S**. The union and intersection of two languages over an alphabet are also languages over the same alphabet. If $u = (u_1 u_2 u_3 \cdots u_m)$ and $v = (v_1 v_2 \cdots v_n)$ are two strings of lengths m and n, respectively in S^* then the **concatenation** of u and v is the string uv in S^* of length $m + n$ defined as $uv = (u_1 u_2 u_3 \cdots u_m v_1 v_2 \cdots v_n)$. The mapping $c: S^* \times S^* \rightarrow S^*$ defined by $c(u, v) = uv$ where uv is the concatenation of u and v is a binary operator on S^*.

Relations

We conclude this section with a brief comment on the concept of a "relation," which is more general than that of a function. If A and B are two sets, any subset of $A \times B$ is called a **relation from A to B**. For example, if $A = \{a, b, c\}$ and $B = \{1, 2, 3, 4\}$, then $R = \{(a, 2), (a, 3), (b, 4), (c, 3)\}$ is a relation from A to B. By definition, in each ordered pair in a relation from A to B, the first element is an element in A and the second element is an element in B. A function from A to B therefore defines a special kind of relation R from A to B such that whenever (a, b) and (a, b') are in the relation R, then $b = b'$. In other words, $f: A \rightarrow B$ defines the cartesian product $\{(x, f(x)) : x \text{ is in } A\}$, which is a subset of $A \times B$.

A relation R from a finite set A with m elements to a finite set with n elements can be represented pictorially by a bipartite graph G with m vertices on the left side (corresponding to the *m* elements of *A*) and *n* vertices on the right side (corresponding to the *n* elements of *B*) as in Figure 0.2.1. If (a, p) is an element in the relation *R*, an arrow is drawn from the vertex *a* on the left side to the vertex *p* on the right side. For example, the graph in Figure 0.2.1 represents the relation $R = \{(a, p), (b, p), (c, r)\}$ from the set $A = \{a, b, c\}$ to the set $B = \{p, q, r\}$.

A relation from a set *A* to itself is called a **relation on** *A*. An informative and useful way to represent a relation *R* on a finite set *A* with *n* elements is by drawing a directed graph with *n* vertices representing the *n* elements of the set and drawing an arrow from vertex *u* to vertex *v* if and only if the ordered pair (u, v) is in the relation. If (u, u) is in the relation, a loop from *u* to *u* is drawn. For example, if $R = \{(a, a), (a, b), (b, c), (c, b)\}$ is a relation on the set $A = \{a, b, c\}$, this relation *R* can be represented by the directed graph shown in Figure 0.2.2.

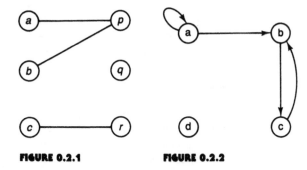

FIGURE 0.2.1   **FIGURE 0.2.2**

A relation *R* on *A* is **reflexive** if (a, a) is an element of *R* for every *a* in *A*, it is **symmetric** if (a, b) is in *R* whenever (b, a) is in *R* and it is **transitive** if (a, c) is in *R* whenever (a, b) and (b, c) are in *R*. A relation *R* on a set is **antisymmetric** if whenever *a* and *b* are distinct, then (a, b) is in the relation *R* only when (b, a) is not in the relation *R*. Finally, the relation *R* is said to have the **comparison property** if either (a, b) or (b, a) is in *R* for all *a* and *b* in *A*.

Suppose that *R* is a relation on a finite set *A* and let *G* be the directed graph that represents this relation. Then *R* is reflexive if and only if there is a loop at every vertex of *G* and *R* is symmetric if and only if whenever there is an arrow from *a* to *b*, there is an arrow from *b* to *a*. Furthermore, *R* is transitive if and only if whenever there is an arrow from *a* to *b* and an arrow from *b* to *c* there is an arrow from *a* to *c*.

Example 0.2.7

Let $A = \{a, b, c\}$ and let *R* be a relation on *A*.

(a) $R = \{(a, b), (b, a), (a, a), (b, b), (b, c), (c, c)\}$ is reflexive because (u, u) is in R for all u in A. Here (a, a), (b, b), and (c, c) are in R. These three elements will represent loops at the three vertices of the corresponding digraph.

(b) $R = \{(a, b), (b, a), (c, c)\}$ is symmetric because whenever (u, v) is in R for any u and any v in A, then (v, u) also is in R. Here both (a, b) and (b, a) as well as (c, c) are in R. In the digraph that represents this relation there will be arrows from a and b and from b to a. There will be a loop at the vertex c.

(c) $R = \{(a, b), (b, c), (a, c), (b, b)\}$ is transitive.

(d) $R = \{(a, c), (b, b), (a, b), (a, a)\}$ is antisymmetric.

(e) If $R = \{(a, c), (b, b), (c, c), (a, b), (c, b)\}$, then R has the comparison property.

Equivalence Relations

A relation S on a set is called an **equivalence relation on** A if S is reflexive, symmetric, and transitive. For example, if $S = \{(a, b) : a, b \text{ are real}, a = b\}$, then S is obviously an equivalence relation on the set of real numbers.

Example 0.2.8

(a) Let $A = \{a, b, c, d, e\}$ and $C(A)$ be a partition of A defined by the class $\{\{a, b\}, \{c, d, e\}\}$. Let R be the set of ordered pairs (x, y) in $A \times A$ such that whenever x is in one of the sets in the partition, then y is also in the same set. Thus in this case $R = \{(a, a), (b, b), (c, c), (d, d), (e, e), (a, b), (b, a), (c, d), (d, c), (c, e), (e, c), (d, d), (d, e)\}$. It is easily verified that R is an equivalence relation. Every partition of a set defines a unique equivalence relation on it.

(b) Conversely, it can easily be established that every equivalence relation on a set defines a partition on the set. If the ordered pair (a, b) belongs to an equivalence relation on a set A, we take both a and b belong to the same subset of A. The class of subsets thus formed constitutes a partition of X. For instance the equivalence relation $R = \{(p, p), (q, q), (p, q), (q, p), (r, r)\}$ defines the partition $\{\{p, q\}, \{r\}\}$ of the set $\{p, q, r\}$.

Equivalence Sets and the Equivalence Class

Let R be an equivalence relation on a set A and let x be any element of A. The **equivalence set** $[x]$ of the element x is the set $\{y : (y, x) \in R\}$. Observe that if $[u]$ and $[v]$ are two distinct equivalent sets, their intersection is empty. For if x is in both $[u]$ and in $[v]$, then because of transitivity (u, v) is in the relation R that implies $[u] = [v]$. The class of distinct equivalent sets of the elements in X is called the **equivalence class** of the relation. An equivalence class of a set is a partition of a set, and vice versa. Thus there is no real distinction between partitions of a set and equivalence classes in the set. In practice, it is almost invariably the case that we use equivalence relations to obtain partitions because it is usually easy to define an equivalence relation on a set.

Partial Orders and Linear Orders

A relation R on A is a **partial order** if it is reflexive, antisymmetric, and transitive. A partial order R that has the comparison property is called a **total** (or **linear**) **order.** A nonempty set A together with a partial order relation P defined on it is called a **partially ordered set** (PO set) and is denoted by (A, P). A partially ordered set (A, P) is called a **totally (linearly) ordered set** or a **chain** if P has the comparison property.

Example 0.2.9

(a) Let A be nonempty set and $P(A)$ its power set. Let R be a relation on $P(A) \times P(A)$ defined by the "set-inclusion" property; that is, if E and F are subsets of A, then (E, F) is in the relation R if E is subset of F. Then R is a partial order on $P(A)$ and $(P(A), R)$ is a partially ordered set. But it is not a linearly ordered set for an arbitrary subset of A need not contain another arbitrary subset of A.

(b) If x and y are two real numbers, we say that (x, y) is an element in the relation S on the set R of real numbers whenever x is less than or equal to y. Then the relation S is a linear order on R.

Example 0.2.10

Let $X = \{1, 2, 3, 4, 5, 6, 7, 8, 9, 10\}$ and S be the relation on X defined as $S = \{(m, n) : m \text{ divides } n\}$. Then S is a partial order on S. The set $A = \{2, 4, 8\}$ is a chain in X, whereas the set $B = \{2, 5, 10\}$ is not a chain since the elements 2 and 5 are not comparable.

Hasse Diagrams of Partially Ordered Sets

Consider the directed graph G that represents a partial order R on a finite set A. Since R is reflexive, there is a loop at each vertex of the graph. Since R is transitive, there is an arc from the vertex u to the vertex v whenever there is an arc from u to w and an arc from w to v. So we can have a simplified pictorial representation of the partial order if we ignore the loops and delete all arrows that are present due to transitivity. Furthermore, if the graphical representation is so oriented that all arrows point in one direction (upward, downward, left to right, or right to left), we can ignore the direction of the arrows as well. The resulting diagram is called a **Hasse diagram** of the partially ordered set.

Example 0.2.11

Let $X = \{1, 2, 3, 4, 5\}$ and $S = \{(1, 1), (1, 2), (1, 3), (1,4), (1, 5),$ $(2, 2), (2, 3), (2, 5), (3, 3), (3, 5), (4, 4), (4, 5), (5, 5)\}$. It can be easily verified that S is a partial order on X. The Hasse diagram that represents S is shown in Figure 0.2.3.

Maximal and Minimal Elements

An element u in a partially ordered set A with a partial order R is called a **maximal element** in the set if whenever (u, x) is in R, then $x = u$. Similarly, an element v in A is a **minimal element** if whenever (x, v) is in R, then $x = v$.

Example 0.2.12

Let $X = \{2, 3, 4, 5, 8, 12, 24, 25\}$ and let R be the partial order on X defined by $R = \{(m, n) : m$ divides $n\}$. Then 2 is a minimal element of R because no element in X divides 2. Similarly, 3 and 5 are also minimal elements of R. Similarly, 24 is a maximal element because there is no number in X that is divisible by 24. Another maximal element in X is 25.

The minimal and maximal elements of a partial order can easily be spotted using its Hasse diagram, in which the minimal elements will be on the bottom and the maximal elements will be at the top if all the arrows are drawn upward. See Figure 0.2.4, representing the Hasse diagram of

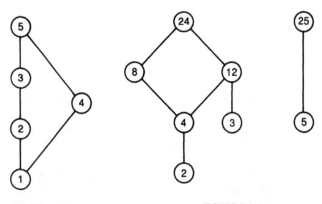

FIGURE 0.2.3   **FIGURE 0.2.4**

Example 0.2.12, in which the vertices representing 24 and 25 are at the top and the vertices representing 2, 3, and 5 are at the bottom.

Example 0.2.13

$P(A)$ is the partially ordered set of all subsets of A with the partial order defined by set inclusion, and in this PO set, A is the only maximal element and the empty set is the only minimal element.

A partially ordered set may have more than one maximal (or minimal), as we saw in Example 0.2.12. There are partially ordered sets with no maximal or minimal elements. Consider the relation $S = \{(x, y) : x, y \text{ are integers}, x \leq y\}$. Then S is no doubt a partial order on the set Z of integers, but this PO set has no maximal or minimal element.

Maximum (Greatest) and Minimum (Least) Elements

An element M in a partially ordered set A with a partial order S is called a **maximum** (or **greatest element**) in A if $(x, M) \in S$ for every x in the set A. Similarly, an element m is a **minimum** (or **least element**) if $(m, x) \in S$ for every x the set A.

[One should be very careful in distinguishing (1) between a maximal element and a maximum element and (2) between a minimal element and a minimum element. If an element is a maximum or a minimum, all elements in the set must be comparable to it. Of course, if a maximum element exists, it is no doubt a maximal element. Similarly, if a minimum element exists, it is a minimal element. The converse implications are not necessarily true, as can be seen from the Hasse diagrams in Example 0.2.14. In a multiparty government, each party leader can be considered as a maximal element, whereas in a single-party system the unique party leader is both maximum and maximal.]

Example 0.2.14

Let $A = \{1, 2, 3, 4\}$ and consider the four partial orders on A with Hasse diagrams as in Figure 0.2.5. In part (a), 4 is the greatest element and 1 is the least element. In (b), 4 is the greatest element and the minimal elements are 1 and 2. There is no least element in (b). In (c), 1 is a least element. There is no greatest element here; but 2, 3, and 4 are maximal elements. There are no greatest or least elements in (d). The elements 1 and 2 are minimal and the elements 3 and 4 are maximal.

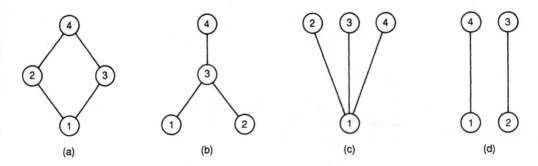

(a)   (b)   (c)   (d)

FIGURE 0.2.5

Example 0.2.15

In each of the following sets of positive integers, the ordered pair (m, n) is in a relation S if m divides n.

(a) $A = \{2, 4, 6, 8\}$. Here 2 is the least element. There is no greatest element. The maximal elements are 6 and 8.

(b) $A = \{2, 3, 4, 12\}$. The greatest element is 12. There is no least element. The minimal elements are 2 and 3.

(c) $A = \{2, 4, 8, 16\}$. The greatest is 16 and the least is 2.

(d) $A = \{2, 3, 4, 6\}$. Here 2 and 3 are minimal; 4 and 6 are maximal.

Well-Ordered Sets

A partially ordered set A in which every subset B has a least element m in B is called a **well-ordered set.** For example, if N is the set of all positive integers, and if we say that (a, b) is in S whenever a is less than or equal to b, then S is a partial order on N. Let B be any subset of N. Obviously, the smallest integer in B is the least element of B. Thus every subset of N has a least element in it. So N is a well-ordered set. The set A of real numbers in an interval is not a well-ordered set under the relation S, where (x, y) in S means that x is less than or equal to y. A subset of a well-ordered set is well-ordered. A well-ordered set A is linearly ordered because any set of two elements in A has a first element and therefore the relation has the comparison property.

Zorn's Lemma

Let B be a subset of a partially ordered set A with the partial order relation S. An element u in A is called an **upper bound of B** if (x, u) is in S for all x in B. Observe that u need not be in B. If there exists an upper bound for B, we say that B has an upper bound. An arbitrary subset of a PO set need not have an upper bound.

One of the most important and exceedingly powerful tools of mathematics is **Zorn's lemma,** which asserts that if A is a partially ordered set in which every chain has an upper bound, then A has a maximal element. This lemma cannot be "proved" in the usual sense of the term. However, it can be shown that it is logically equivalent to the celebrated **axiom of choice,** which lies at the very foundation of set theory. Thus Zorn's lemma is assumed as an axiom of logic and set theory. Many important existence theorems are proved by invoking Zorn's lemma. The axiom of choice is also logically equivalent to the well-ordering theorem of Zermelo: Every set can be well-ordered. For a proof of Zorn's lemma, using the axiom of choice, see the book *Naive Set Theory* by P. R. Halmos.

0.3 INDUCTIVE PROOFS AND RECURSIVE DEFINITIONS

One of the most useful, elegant, and simple proof techniques in mathematics in general and in discrete mathematics in particular is the technique known as *mathematical induction*, which is essentially an "algorithmic proof procedure." The origins of this technique can be traced to the days of the classical Greek period. But the term *induction* was coined by De Morgan only in the nineteenth century.

The Principle of Mathematical Induction (Weak Form)

This principle is stated as follows: Suppose that $P(n)$ is a statement about the natural number n and q is a fixed natural number. Then an induction proof that $P(n)$ is true for all $n \geq q$ requires two steps:

1. *Basis step*: Verify that $P(q)$ is true.
2. *Induction step*: Verify that if k is *any* positive integer greater than or equal to q, then $P(k + 1)$ is true whenever $P(k)$ is true.

Here $P(n)$ is called the **inductive hypothesis.** When we complete both steps of a proof by mathematical induction, it is obvious that we have proved that the statement $P(n)$ is true for all positive integers $n \geq q$. [A formal proof is along the following lines: Suppose that $P(n)$ is not true for all $n \geq q$. Then there is at least one positive integer $k \geq q$ such that $P(k)$ is not true. So the set D of positive integers r for which $P(r)$ is not true is nonempty, and therefore this set has a unique least element because any set of positive integers is well-ordered. Let t be the first element of this set. Of course, $t > q$. Since t is an integer, $t - 1$ also is an integer which is not in D. So $P(t - 1)$ is true, which implies by the induction step that $P(t)$ is true. This is a contradiction.]

This version of induction given above is called the *weak form* because the induction step assumes that $P(n)$ is true in exactly one case. A *strong form* of induction is discussed later in this section.

Example 0.3.1

Prove that $1 + 2 + 3 + \cdots + n = n(n + 1)/2$ for all natural numbers.

Proof (By Mathematical Induction). Let $P(n)$ be the statement that $(1 + 2 + 3 + \cdots + n)$ is equal to $n(n + 1)/2$. The aim is to prove that $P(n)$ is true for all n.

The basis step: We have to verify that $P(1)$ is true. Here $P(1)$ is the statement that 1 is equal to $1(1 + 1)/2$. This is true. So $P(1)$ is true.

The induction step: We have to verify that if $P(k)$ is true, then $P(k + 1)$ is true. Now $P(k)$ is the statement that $1 + 2 + \cdots + k$ is equal to $k(k + 1)/2$ and $P(k + 1)$ is the statement that $1 + 2 + \cdots + (k + 1)$ is equal to $(k + 1)(k + 2)/2$. If $P(k)$ is true, $1 + 2 + \cdots + k = k(k + 1)/2$. Thus $1 + 2 + \cdots + k + (k + 1) = k(k + 1)/2 + (k + 1)$, which implies that $1 + 2 + \cdots + (k + 1) = (k + 1) \cdot (k + 2)/2$. So $P(k + 1)$ is true whenever $P(k)$ is true.

[One has to be very careful in using the induction method to prove theorems. The meaning of the induction step is very precise: If $P(q)$ is true, $P(q + 1)$ is true. If $P(q + 1)$ is true, then $P(q + 2)$ is true, and so on. In other words, the induction should hold for *every* k greater than or equal to q. An erroneous "proof" justifying the statement that all roses are of the same color is as follows. $P(n)$ is the proposition that all roses in any collection of n roses are of the same color. Our aim is to show that $P(n)$ is true for all n. Obviously, $P(1)$ is true. Suppose that $P(k)$ is true for some positive integer k. So all the roses in any collection of k roses are of the same colors. Now consider any arbitrary collection C of $(k + 1)$ roses. Label these roses as r_i ($i = 1, 2, 3, \ldots, k + 1$). Let A be the set $\{r_i : i = 1, 2, \ldots, k\}$ and B be the set $\{r_i : i = 2, 3, \ldots, k, k + 1\}$. Both A and B have exactly k roses. So the roses in A are all of the same color. Similarly, the roses in B also are of the same color. The rose labeled r_k is in both A and B. So all the $k + 1$ roses under consideration are of the same color. So if $P(k)$ is true, then $P(k + 1)$ is true. Can we therefore conclude that $P(n)$ is true for all n? The answer is "no" because when the set C has two elements, the sets A and B are disjoint. So if $P(1)$ is true, $P(2)$ need not be true.]

Example 0.3.2

Use induction to prove that the sum of the first n odd positive integers is n^2.

Proof.   Let $P(n)$ be the statement that the sum of the first n odd positive integers is n^2. The aim is to prove that $P(n)$ is true for every n.
  The basis step: $P(1)$ is true because $1 = 1^2$.
  The induction step: We have to verify that $P(k + 1)$ is true whenever $P(k)$ is true for any positive integer k. Since $P(k)$ is true, $1 + 3 + 5 + \cdots + (2k - 1) = k^2$. Consequently,

$$1 + 3 + 5 + \cdots + (2k - 1) + (2k + 1) = k^2 + (2k + 1) = (k + 1)^2$$

So $P(k + 1)$ is true. Thus $P(n)$ is true for every n.

Example 0.3.3

Use induction to prove that $n < 2^n$ for any positive integer n.

Proof.   Let $P(n)$ be the statement that $n < 2^n$ for the positive integer n. We have to prove that $P(n)$ is true for any positive integer.
  The basis step: $P(1)$ is true since $1 < 2$.
  The induction step: Suppose that $P(k)$ is true for an arbitrary positive integer, implying that $k < 2^k$. Then $k + 1 < 2^k + 1 < 2^k + 2^k$. Hence $k + 1 < 2^{k+1}$, which implies that $P(k + 1)$ is true. So $P(n)$ is true for all positive integers n.

The Principle of Mathematical Induction (Strong Form)

Suppose that $P(n)$ is a statement about the natural number n and q is a fixed natural number. Then an induction proof that $P(n)$ is true for all $n \geq q$ requires two steps:

1. *Basis step*: Verify that $P(q)$ is true.
2. *Induction step*: Verify that if $k \geq q$ and if $P(q), P(q + 1), P(q + 2), \ldots, P(k)$ are true, then $P(k + 1)$ is true.

  (This version of the induction principle is "strong" in the sense that the induction step here has more information than that of the induction step in the "weak" version. As in the previous case, it can easily be shown that the strong version is also a consequence of the fact that any set of natural numbers is well-ordered. So to prove a theorem using mathematical induction, one can use either version of mathematical induction. In some cases it is more convenient to use the strong form, as seen in the next example. The strong form of induction is also known as **complete induction**.)

Example 0.3.4

Prove that any natural number greater than 1 can be factored as a product of prime numbers.

Proof (By Complete Mathematical Induction).   Let $P(n)$ be the statement that when n is a natural number greater than 1, then n can be factored as a product of prime numbers. The aim is to prove that $P(n)$ is true for all n.

The basis step: $P(2)$ is the statement that 2 can be factored as a product of primes. Obviously, $P(2)$ is true.

The induction step: Suppose that $P(2), P(3), \ldots, P(k)$ is true. We have to verify that $P(k + 1)$ is true. Now $P(k + 1)$ is certainly true when $k + 1$ is a prime number. If $k + 1$ is not a prime number, we can always find two positive integers m and n such that $k + 1 = mn$, where both m and n are less than k. By the induction step, both m and n can be expressed as the product of prime numbers. So $k + 1$ also can be factored as a product of prime numbers. Thus $P(k + 1)$ is true.

Recursive Definitions of Sets

The basic idea underlying the principle of induction is as follows. Once we describe the initial stage in some process and if we are able to describe any subsequent stage in terms of the previous stages, we are in a position to describe the entire process completely at all stages. The parallel concept in computer science is **recursion,** where we tend to think of the process in the opposite direction. Informally, this is the process of solving a large problem by decomposing it into one or more subproblems such that each such subproblem is identical in structure to the original problem but more or less simpler to solve. So in both situations, one must (1) decide a set of simple cases for which the proof or computation is easily handled, and (2) obtain an appropriate rule that can be applied repeatedly until the end. This concept underlying both induction and recursion can be used to justify the definition of some collection of objects in stages. Such a description is called aptly an **inductive** or **recursive definition.**

A recursive definition of a set consists of three parts:

1. *Basis part*: This part tells us that certain elements belong to the set we are going to define.
2. *Inductive (recursive) part*: This part tells us to use the elements currently in the set to obtain more objects that can be included in the set.
3. *Closure part*: This part tells that the only elements in the set are those obtained by (1) and (2).

Example 0.3.5

To define the set A of positive integers divisible by the number 5 recursively, we have the recursive definition consisting of the following three parts:

   (a) 5 is an element of A.

   (b) If n is an element of A, then $n + 5$ also is an element of A.

   (c) An object is in A if and only if it is obtained by a repeated application of (a) and (b).

Recursive Definitions of Functions

Suppose that (1) each element i in the set $S = \{0, 1, 2, \ldots, k\}$ of the first $k + 1$ consecutive integers is assigned a real number r_i, and (2) if n is any integer greater than k, there is a rule f for defining a real number $f(n)$ which can be expressed uniquely in terms of some or all the terms from the set $\{f(n - 1),$ $f(n - 2), \ldots, f(n - k), f(n - k - 1)\}$. If we now define $f(i) = r_i$ for each i in S, the rule f is a function whose domain is the set of nonnegative integers. A function defined by this method is called a **recursively defined function.** The rule that defines $f(n)$ in terms of the preceding values $f(i)$ is called a **recurrence relation.** The values $f(0), f(1), f(2), \ldots, f(k)$ are called the **initial values** of the recurrence relation.

   We use the induction principle (strong form) to show that this definition does not violate the true definition of a function, [i.e., $f(n)$ is unique for every nonnegative integer n].

THEOREM 0.3.1

If f is recursively defined, then $f(n)$ is unique for every positive integer n.

Proof:

$P(n)$ is the proposition that $f(n)$ is unique for every n.

The basis step: We assume that $f(i) = r_i$ is unique when $i = 0, 1, 2, \ldots,$ k. So $P(0), P(1), \ldots, P(k)$ is true.

The induction step: $f(k + 1)$ is expressed uniquely in terms of these $k + 1$ numbers. So $P(k + 1)$ is true.   ◇

Example 0.3.6

Suppose that N is the set of ll nonnegative integers and R is the set of all real numbers.

   (a) The function $f: N \to R$ that defines the sequence $f(n) = 3^n$ can be recursively defined as $f(0) = 1$ and $f(n) = 3f(n - 1)$ when $n > 0$.

(b) The (factorial) function $f: N \rightarrow R$, where $f(n) = n!$, can be recursively defined as $f(0) = 1$ and $f(n) = nf(n - 1)$ when $n > 0$.

(c) The Fibonacci sequence $f: N \rightarrow R$ defined recursively by the relation $f(n) = f(n - 1) + f(n - 2)$ when $n > 1$, with the initial values $f(0) = 0$ and $f(1) = 1$ gives the sequence $\{0, 1, 1, 2, 3, 5, 8, 13, \ldots\}$.

It is important that in a recursive definition, the set of integers in the basis step constituting the initial conditions is a *consecutive* set of integers. Otherwise, the function may not be well-defined. Here is a counterexample: $f(n) = 9f(n - 2)$, with the nonconsecutive initial values $f(0) = 6$ and $f(2) = 54$, will yield $f(n) = 2 \cdot 3^{n+1}$ as well as $f(n) = 3 \cdot 3^n + 3 \cdot (-3)^n$.

0.4 *THE LANGUAGE OF LOGIC*

At an introductory level, mathematical logic is very similar to set theory. Instead of sets, in logic we have **propositions.** A proposition is a statement that is either **true** or **false** but not both. The **truth value** of a proposition p is T (or 1) if p is true; otherwise, the truth value is F (or 0). Consider the following five sentences:

1. p: $3 + 2 = 5$
2. q: $3 + 2 = 6$
3. r: Is it 3 or 2?
4. s: Take 3
5. t: $x + 2 = 5$

Here p is a true proposition and q is a false proposition. Both r and s are not propositions. t is a proposition, but it is neither true nor false since x is unknown.

In set theory we have the intersection and union of two sets and the complement of a set in a certain universal set. The analogous concepts in logic are the three logical operations: the conjunction of two propositions, the disjunction of two propositions, and the negation of a proposition.

The **conjunction** of two propositions p and q is a proposition that is true if and only if both p and q are true propositions. The conjunction of p and q is called "p and q" and is denoted by $p \wedge q$. The **disjunction** of two propositions p and q is a proposition that is false if and only if both p and q are false propositions. The disjunction of p and q is called "p or q" and is denoted by $p \vee q$. The **exclusive disjunction** of two propositions p and q is a proposition that is true if and only exactly one of the two is true and is denoted by $p \oplus q$. Finally, the **negation** of a proposition p is a proposition p' which is true if and only if p is false.

A **truth table** displays the relationships between the truth values of propositions. The following table displays the truth values of conjunction, disjunction, exclusive disjunction, and negation of two propositions p and q:

p	q	Conjunction	Disjunction	Exclusive disjunction	Negation of p
T	T	T	T	F	F
T	F	F	T	T	F
F	T	F	T	T	T
F	F	F	F	F	T

Propositions that can be obtained by the combination of other propositions are known as **compound propositions.** For example, if p, q, and r are three propositions, then "$(p'$ and $q)$ or (q)" is a compound proposition. A proposition that is not a combination of other propositions is called an **atomic proposition.**

The Implication Operation

There is another important way to construct a compound proposition from two propositions p and q. This is the **implication** proposition: **"if p, then q"** (or "p **implies** q") which is denoted by $p \rightarrow q$. We define this compound proposition to be false if and only when p is true and q is false. In all other cases the compound proposition $p \rightarrow q$ is true. In this case we say that the proposition p is a **sufficient condition** for the proposition q and q is a **necessary condition** for p. Here p is called the **hypothesis** (or **antecedent** or **premise**) and q is called the **consequence** (or **conclusion**). Obviously, the compound proposition $p \rightarrow q$ and the compound proposition $q \rightarrow p$ both cannot be false at the same time. So irrespective of the truth values of p and q, it is always the case that at least one of these two compound propositions $p \rightarrow q$ and $q \rightarrow p$ is always true.

Observe that the mathematical concept of implication is more general than the concept of implication regarding statements in the languages we use to communicate in our daily lives. In a general mathematical setting there is no cause-and-effect relationship between the truth value of the hypothesis and the truth value of the conclusion. For example, if p is the proposition that "it is raining today" and q is the proposition that "London is the capital of England," then the implication proposition $p \rightarrow q$ is true whether or not p is true. The implication that "if it rains today, then $3 + 4 = 8$" is a true proposition if it does not rain today. On the other hand, when I make the statement that "if it rains like this, I will not go fishing this afternoon," this statement is an implication proposition in which there is a definite causal relation between the hypothesis and the conclusion.

It is also important to mention in this context that the implication proposition in many programming languages has a different truth value. If a line in a program says "if $n < 30$ then S," then when the execution of the program reaches this line, the segment S is executed if $n < 30$ and not executed otherwise.

The compound proposition "p **if and only if** q," which is denoted by $p \leftrightarrow q$, is the conjunction of the compound proposition $p \rightarrow q$ and the compound proposition $q \rightarrow p$. The compound proposition $p \leftrightarrow q$ is true when both $p \rightarrow q$ and $q \rightarrow p$ are true. In this case we say that p is both **necessary and sufficient** for q, and vice versa. Here is the truth table of these implication operations involving two operations p and q:

p	q	$p \rightarrow q$	$q \rightarrow p$	$p \leftrightarrow q$
F	F	T	T	T
F	T	T	F	F
T	F	F	T	F
T	T	T	T	T

If r is the proposition $p \rightarrow q$, the proposition $q \rightarrow p$ is the **converse** of r, the proposition $p' \rightarrow q'$ is the **inverse** of r, and the proposition $q' \rightarrow p'$ is the **contrapositive** of r. If two propositions p and q are such that p is true if and only if q is true, the two propositions p and q are said to be **equivalent.** We write $p = q$ when p and q are equivalent. In other words, two propositions are equivalent if they have the same truth value. For example, the proposition that "Jane will be 18 years old in 1993" is equivalent to the proposition that "Jane was born in 1975." The compound proposition $p \rightarrow q$ is equivalent to its contrapositive proposition $q' \rightarrow p'$ since they both have the same truth value, as can be seen from the following truth table.

p	q	p'	q'	$p \rightarrow q$	$q' \rightarrow p'$
T	T	F	F	T	T
F	T	T	F	T	T
T	F	F	T	F	F
F	F	T	T	T	T

A compound proposition that is always true irrespective of the truth values of its component propositions is called a **tautology.** A compound proposition that is always false is called a **contradiction.** As a simple example, the disjunction of a proposition p and its negation is a tautology, whereas the conjunction of p and its negation is a contradiction. From the table given above we notice that the proposition $p \leftrightarrow q$ is a tautology if and only if $p = q$.

The commutative, associative, and distributive laws involving the conjunction and disjunction operation can easily be verified by constructing the appropriate truth tables. These laws are:

1. *The commutative laws:* $p \wedge q = q \wedge p$ and
$$p \vee q = q \vee p$$
2. *The associative laws:* $p \wedge (q \wedge r) = (p \wedge q) \wedge r$ and
$$p \vee (q \vee r) = (p \vee q) \vee r$$
3. *The distributive laws:* $p \wedge (q \vee r) = (p \wedge q) \vee (p \wedge r)$ and
$$p \vee (q \wedge r) = (p \vee q) \wedge (p \vee r)$$

The truth value of a compound proposition depends on the truth values of its component propositions. The subject of constructing and simplifying compound propositions built from other propositions and obtaining their truth values is called **propositional calculus.** This combination of propositions to yield new propositions bears a strong resemblance to the combination of sets to form new sets. The following theorem is analogous to De Morgan's laws on set theory.

THEOREM 0.4.1

(a) $(p \wedge q)' = (p') \vee (q')$.
(b) $(p \vee q)' = (p') \wedge (q')$.

Proof:

This is left as an exercise.   ◇

The Satisfiability Problem in Logic

The truth table of a compound proposition with n atomic propositions as its components will have 2^n rows. The **satisfiability problem** of a compound proposition p is the problem of (1) finding out whether there exist truth values for the atomic components of p such that p is true, and (2) obtaining the true atomic propositions and the false atomic propositions if they exist which make the compound proposition true. The only known procedure for testing the satisfiability of a proposition with n atomic propositions is building a truth table enumerating all the 2^n possibilities of truth values, and this is a formidable task indeed when n is large.

0.5 NOTES AND REFERENCES

A systematic investigation of the theory of sets began with the contributions of Georg Cantor (1848–1918) in the nineteenth century. Prior to that, set theory and more generally nonnumerical mathematics were not investigated in a formal manner apart from the contributions of George Peacock (1791–1858), Augustus De Morgan (1806–1871), and George Boole (1815–1864). Peacock and De Morgan generalized the usual algebraic operations beyond the realm of numerical mathematics and Boole extended and formalized their contributions in his seminal work entitled "Investigation of the Laws of Thought" in 1854. According to the great twentieth-century philosopher and mathematician Bertrand Russell (1872–1970), it was George Boole who "discovered" pure mathematics.

More on the history and development of set theory can be found in Boyer (1968). For a detailed treatment of set theory, including functions and relations, the books by Halmos (1960) and Stoll (1963) are highly recommended. The technique of proof by induction was explicitly stated and used by Francesco Maurolycus in the sixteenth century when he proved that the sum of the first n odd positive integers is n^2. However, this technique was known to mathematicians as early as the third century B.C. For example, in Euclid's proof that there are an infinite number of primes, this technique was used implicitly.

In the seventeenth century, both Pascal and Fermat used the induction method extensively. The term *mathematical induction* was coined by De Morgan. In the nineteenth century, the principle of induction was investigated in detail by Gottlieb Frege (1848–1925), Giuseppe Peano (1858–1932), and Richard Dedekind (1831–1916). The role of induction in the formal development of mathematics became a primary focus of many mathematical logicians at the beginning of the twentieth century, and two names worth mentioning in this regard are those of Bertrand Russell and Thoralf Skolem (1887–1963). For an interesting survey on the topic of mathematical induction, see the article by Bussey (1917). Golovina and Yaglom (1963), Polya (1954), and Sominskii (1963) are three excellent references in this area. The article by Henkin (1960) is also highly recommended.

The origins of a systematic study of logical reasoning can be traced to Aristotle, who lived in the fourth century B.C. It was not until the seventeenth century, however, that symbols were used in the study of logic. The pioneering work in symbolic logic was done by Gottfried Leibniz (1646–1716). No major developments took place until George Boole published his outstanding work mentioned earlier. Since then Bertrand Russell and Alfred North Whitehead (1861–1947) contributed considerably to the development of logic and the field of mathematical logic emerged when the discovery of certain paradoxes led to an extensive examination of the place of logic, proof, and set theory in the foundations of mathematics.

0.6 EXERCISES

0.1. Let $A = \{3, 5, 7, 9\}$, $B = \{2, 3, 5, 6, 7\}$, and $C = \{2, 4, 6, 8\}$ be all subsets of the universe $X = \{2, 3, 4, 5, 6, 7, 8, 9\}$. Find (a) the union of A and B, (b) the intersection of B and C, (c) $B - A$, (d) $A - B$, (e) the absolute complement

C' of the set C, **(f)** the absolute complement of X, **(g)** the absolute complement of the empty set.

0.2. Let A, B, C be as in Problem 0.1. Find the following sets: **(a)** $(A \cup B) - C$, **(b)** the intersection of $(A \cup B)'$ and $(B \cup C)'$, **(c)** $(A \cup C) - (C - A)'$

0.3. Which of the following sets are equal? **(a)** $\{a, b, c, c\}$, **(b)** $\{a, b, a, b, c\}$, **(c)** $\{a, b, b, c, d\}$

0.4. Which of the following sets are equal?
  (a) $\{t : t$ is a root of $x^2 - 6x + 8 = 0\}$
  (b) $\{y : y$ is a real number in the closed interval $[2, 3]\}$
  (c) $\{4, 2, 5, 4\}$
  (d) $\{4, 5, 7, 2\} - \{5, 7\}$
  (e) $\{q : q$ is either the number of sides of a rectangle or the number of digits in any integer between 11 and 99$\}$

0.5. If $A = \{3, 4\}$ and $B = \{p, q, r\}$, list all the elements of **(a)** $A \times A$, **(b)** $A \times B$, **(c)** $B \times A$ and **(d)** $B \times B$.

0.6. Let A and B be as in Problem 0.5. List all the elements of **(a)** $A \times A \times A$ and **(b)** $A \times A \times B$.

0.7. Let A and B be as in Problem 0.5. List all the elements of **(a)** $A \cup (B \times A)$ and **(b)** $(A \times A) \cup (B \times A)$.

0.8. List all the sets in the power set of the following sets: **(a)** $\{a, b\}$, **(b)** $\{a, b, c\}$, **(c)** $\{\phi, 0, \{0\}\}$

0.9. List all partitions of the following sets: **(a)** $\{a\}$, **(b)** $\{a, b\}$, **(c)** $\{a, b, c\}$

0.10. Determine whether each of the following statements is true in the case of three arbitrary sets P, Q, R.
  (a) If P is an element of Q and if Q is a subset of R, then P is an element of R.
  (b) If P is an element of Q and if Q is a subset of R, then P also is a subset of R.
  (c) If P is a subset of Q and Q is an element of R, then P is an element of R.
  (d) If P is a subset of Q and Q is an element of R, then P is a subset of R.

0.11. Prove the following assertions involving three arbitrary sets P, Q, and R.
  (a) $(P - Q) - R = P - (Q \cup R)$
  (b) $(P - Q) - R = (P - R) - Q$
  (c) $(P - Q) - R = (P - R) - (Q - R)$

0.12. Two sets A and B are such that their union and their intersection are equal. What can we say about A and B?

0.13. Suppose that A is a subset of B and C is a subset of D.
  (a) Is it true that $(A \cup C)$ is a subset of $(B \cup D)$?
  (b) Is it true that the intersection of A and C is a subset of the intersection of B and D?

0.14. What can we say about two sets P and Q if $P - Q$ is equal to $Q - P$?

0.15. Prove that A and B are nonempty sets such that if $A \times B$ and $B \times A$ are equal, then $A = B$.

0.16. What is the cardinality of $P \times Q$ if the cardinality of P is p and the cardinality of Q is q?

0.17. Prove that the intersection of the powerset of A and the powerset of B is the powerset of the intersection of A and B, where A and B are two arbitrary sets.

0.18. What can we say if the operation "intersection" is replaced by the operation "union" in Problem 0.17?

0.19. What is the cardinality of the powerset of the empty set?

0.20. If the powerset of A is equal to the powerset of B, does it follow that A and B are equal?

0.21. The **symmetric difference** of two sets A and B is the set containing those elements of either A or B but not both A and B and is denoted by $A \oplus B$. Prove: $A \oplus B = (A - B) \cup (B - A) = (A \cup B) - (A \cap B)$.

0.22. Draw a Venn diagram to represent the symmetric difference of two sets.

0.23. How many distinct regions are there in a Venn diagram that represents three sets in a universal set such that no intersection is empty?

0.24. If the symmetric difference of two sets A and B is equal to the set A, what can we say about A and B?

0.25. If A and B are two arbitrary sets, under what conditions can we conclude that the symmetric difference of $(A - B)$ and $(B - A)$ is the empty set?

0.26. Using Venn diagrams, investigate whether the following statements are true or false.
   (a) $A \oplus (B \cap C) = (A \oplus B) \cap (A \oplus C)$
   (b) $A \oplus (B \cup C) = (A \oplus B) \cup (A \oplus C)$
   (c) $A \oplus (B \oplus C) = (A \oplus B) \oplus C$
   (d) $A \cap (B \oplus C) = (A \cap B) \oplus (A \cap C)$
   (e) $A \cup (B \oplus C) = (A \cup B) \oplus (A \cup C)$

0.27. If the symmetric difference of A and B is equal to the symmetric difference of A and C, is it necessary that $B = C$?

0.28. Prove the following assertions involving three arbitrary sets A, B, and C:
   (a) $A \times (B \cap C) = (A \times B) \cap (A \times C)$
   (b) $A \times (B \cup C) = (A \times B) \cup (A \times C)$
   (c) $(A \cap B) \times C = (A \times C) \cap (B \times C)$
   (d) $(A \cup B) \times C = (A \times C) \cup (B \times C)$

0.29. Let R be the set of all real numbers and $f: R \to R$ defined by $f(x) = x^2$.
   (a) What are the domain, codomain, and range of this function?
   (b) Is f an injection?
   (c) Is f a surjection?
   (d) Find the set of all preimages of 4.
   (e) Find the inverse image of the set $\{t : 1 \leq t \leq 4\}$.

0.30. If R is the set of all real numbers, explain why $F(x) = 1/(x - 2)$ and $F(x) = $ (the square root of x) are not functions from R to R.

0.31. If N is the set of all natural numbers and if $f: N \rightarrow N$ is defined by $f(n) = 2n + 5$, show that f is an injection and find the inverse function. Is f a surjection? Is the inverse function a surjection?

0.32 Suppose that $f(x) = x^2 - 4$, where x is a real number. Find the images of the following sets: **(a)** $\{-4, 4, 5\}$, **(b)** $\{4, 5\}$ **(c)** $\{t : t$ is a real number greater than or equal to zero$\}$.

0.33. Let $A = \{a, b, c, d\}$ and $B = \{p, q, r\}$.
   (a) Find the number of functions from A to B.
   (b) Find the number of injections from A to B.
   (c) Find the number of surjections from A to B.
   (d) Find the number of functions such that a is mapped into p and b is mapped into q.

0.34. If N is the set of all natural numbers, give an example of a function from N to N that is **(a)** an injection but not a surjection, **(b)** a surjection but not an injection.

0.35. Find the domain and range of the function that assigns **(a)** each integer its last digit, **(b)** each integer the number of digits in it.

0.36. Give an example of a function f from the set of real numbers to the set of real numbers such that **(a)** f is both an injection as well as a surjection, **(b)** f is neither an injection nor a surjection.

0.37. Suppose that $X = \{p, q, r\}$, $Y = \{a, b, c, d\}$, and $Z = \{1, 2, 3, 4\}$. Let $g: X \rightarrow Y$ be defined by the set of ordered pairs $\{(p, a), (q, b), (r, c)\}$ and $f: Y \rightarrow Z$ be defined by the set of ordered pairs $\{(a, 1), (b, 1), (c, 2), (d, 3)\}$. Write the composite function $f \circ g$ as a set of ordered pairs.

0.38. If $A = \{p, q, r\}$ and $f: A \rightarrow A$ is defined by $f(p) = q$, $f(q) = p$, and $f(r) = q$, describe f and $f \circ f$ as sets of ordered pairs.

0.39. Let A and f be as in Problem 0.38. Define $f^n = f \circ f \circ f \circ \cdots \circ f$ as the n-fold composition of f with itself. Describe f^n as a set of ordered pairs when n is odd and when n is even.

0.40. Show that the set of all positive integers is equivalent to the set of all positive even integers.

0.41. Let $f: B \rightarrow C$ and $g: A \rightarrow B$. Prove the following:
   (a) If f and g are injections, then $f \circ g$ is an injection.
   (b) If f and g are surjections, then $f \circ g$ is a surjection.

0.42. Let f and g be as in Problem 0.41.
   (a) Suppose that $f \circ g$ is an injection. Is it necessary that f be an injection? Is it necessary that g be an injection?
   (b) Suppose that $f \circ g$ is a surjection. Is it necessary that f be a surjection? Is it necessary that g be a surjection?

0.43. If $f(x) = ax + b$ and $g(x) = cx + d$ and $f \circ g = g \circ f$, find an equation relating a, b, c, and d.

0.44. Suppose that $f: X \rightarrow Y$ and A and B are subsets of X. Then prove: **(a)** $f(A \cup B) = f(A) \cup f(B)$ and **(b)** $f(A \cap B)$ is a subset of the intersection of $f(A)$ and $f(B)$.

0.45. Show that if $f: X \rightarrow Y$ is an injection, then $f(A \cap B) = f(A) \cap f(B)$ for all subsets A and B of X.

0.46. Show that there is an injection from A to B if and only if there is a surjection from B to A.

0.47. Suppose that $f: A \rightarrow B$ where A and B are two finite sets with the same cardinality. Prove that f is an injection if and only if f is a surjection.

0.48. Let A be a subset of a universal set X. The **characteristic function** f_A of A is the function from X to the set $\{0, 1\}$ such that the image of every element in A is 1 and the image of every element not in A is 0. Suppose that A and B are two subsets of X. Prove the following for all x in X.
(a) $f_{A \cap B}(x) = f_A(x) \cdot f_B(x)$ for all x in X
(b) $f_{A \cup B}(x) = f_A(x) + f_B(x) - f_A(x) \cdot f_B(x)$
(c) $f_A(x) + f_{A'}(x) = 1$
(d) If C is the symmetric difference of A and B, then $f_C(x) = f_A(x) + f_B(x) - 2f_A(x) \cdot f_B(x)$.

0.49. Let $S = \{0, 1\}$ and let S_n be the set of all strings of length in S. If u and v are two strings in S_n, we compare them place by place and define the **Hamming distance** $H(u, v)$ between u and v to be the number of places where they differ. Find the Hamming distance between u and v if **(a)** $u = 101100$ and $v = 111011$ **(b)** $u = 01010$ and $v = 11001$.

0.50. Suppose that S, S_n, and $H(u, v)$ are as in Problem 0.49. The function $H: S_n \times S_n \rightarrow N$ (where N is the set of all nonnegative integers) which maps the ordered pair (u, v) into $H(u, v)$ is the **Hamming distance function.** Show that for all u, v, and w in S_n, the function H satisfies the following **metric axioms: (a)** $H(u, v)$ is nonnegative, **(b)** $H(u, v) = 0$ if and only if $u = v$, **(c)** $H(u, v) = H(v, u)$, and **(d)** $H(u, v) \leqslant H(u, w) + H(w, v)$.

0.51. Let $A = \{1, 2, 3, 4, 5\}$, $B = \{a, b, c, d\}$, and R be the relation $\{(1, a), (1, b), (3, c), (4, d), (5, d), (5, c)\}$. Represent this relation by a bipartite graph and draw the appropriate arrows.

0.52. Give an example of a relation on a set that is **(a)** both symmetric and antisymmetric, **(b)** neither symmetric nor antisymmetric.

0.53. Let $A = \{a, b, c, d\}$. Draw the diagraph corresponding to each of the following relations on A and decide whether each relation is reflexive, symmetric, transitive, and antisymmetric. Examine whether the comparison property holds in any of these relations.
(a) $R = \{(b, b), (b, c), (b, d), (c, b), (c, c), (c, d)\}$
(b) $R = \{(a, b), (b, a)\}$
(c) $R = \{(a, a), (b, b), (c, c), (d, d)\}$
(d) $R = \{(a, a), (b, b), (c, c), (d, d), (a, b), (b, a)\}$
(e) $R = \{(a, c), (a, d), (b, c), (b, d), (c, a), (c, d)\}$
(f) $R = \{(a, b), (b, c), (c, d)\}$

0.54. Let R be a relation from A to B and S be a relation from B to C. Then the **composite**

relation $S \circ R$ of R and S is the relation consisting of all ordered pairs of the form (a, c), where (a, b) is in R and (b, c) is in S. If $A = \{p, q, r, s\}$, $B = \{a, b\}$, $C = \{1, 2, 3, 4\}$, $R = \{(p, a), (p, b), (q, b), (r, a) (s, a)\}$ and $S = \{(a, 1), (a, 2), (b, 4)\}$, find $S \circ R$.

0.55. Let R be a relation on the set A. The relation R^2 on A is defined as $R \circ R$. Show that $R^2 \circ R$ is equal $R \circ R^2$. Thus R^3 is the composite of R^2 and R. More generally, the nth power R^n of the relation is the composite of R^{n-1} and R. If $R = \{(a, a), (a, b), (b, a), (c, b), (c, d)\}$, find the second and third powers of R.

0.56. Prove that if a relation on a set is reflexive, then any power of that relation is reflexive.

0.57. Prove that if a relation R on a set is reflexive and transitive, then $R^n = R$ for all positive integers n.

0.58. Let R be a relation from A to B. The **inverse relation** R^{-1} from B to A is the set of all ordered pairs of the form (b, a), where (a, b) is in R. Show that a relation R on a set is symmetric if and only if R and its inverse are equal.

0.59. Show that a relation on a set is reflexive if and only if its inverse relation is reflexive.

0.60. Prove that a relation R on a set A is antisymmetric if and only if the intersection of R and its inverse is a subset of the **diagonal relation** $D = \{(x, x) : x \in A\}$.

0.61. Let R be the relation on the set $A = \{1, 2, 3, 4, 5, 6, 7\}$ defined by the rule $(a, b) \in R$ if the integer $(a - b)$ is divisible by 4. List the elements of R and its inverse.

0.62. Let R be the relation on the set N of all positive integers defined by $(a, b) \in R$ if b is divisible by a. Determine whether R is reflexive, symmetric, antisymmetric, or transitive.

0.63. Let N be the set of all positive integers and let R be the relation on $N \times N$ defined by $((a, b), (c, d))$ is in R if $a \le c$ and $b \le d$. Determine whether R is reflexive, symmetric, antisymmetric, or transitive.

0.64. Which of the following relations on the set $\{1, 2, 3, 4\}$ are equivalence relations? If the relation is an equivalence relation, list the corresponding partition (equivalence class).
(a) $\{(1, 1), (2, 2), (3, 3), (4, 4), (1, 3), (3, 1)\}$
(b) $\{(1, 0), (2, 2), (3, 3), (4, 4)\}$
(c) $\{(1, 1), (2, 2), (1, 2), (2, 1), (3, 3), (4, 4)\}$

0.65. Let $R = \{(x, y) : x$ and y are real numbers and $x - y$ is an integer$\}$. Show that R is an equivalence relation on the set of real numbers.

0.66. Let a be an integer and m be a positive integer. We denote by a (**mod** m) the remainder when a is divided by m. If a and b are two integers, we say that a **is congruent to b modulo** m if m divides $a - b$. The notation $a \equiv b$ (**mod** m) is used to indicate that a is congruent to b modulo m. Of course, if a is congruent to b modulo m, then b is congruent to a modulo m. Prove that:
(a) a (mod m) $= b$ (mod m) if and only if $a \equiv b$ (mod m).
(b) $a \equiv b$ (mod m) if and only if there exists an integer k such that $a = b + km$.

(c) If $a \equiv b \pmod{m}$ and $c \equiv d \pmod{m}$, then $a + c \equiv (b + d) \pmod{m}$ and $ac \equiv bd \pmod{m}$.

0.67. Let Z be the set of all integers and let m be any positive integer greater than 1. Show that the relation R on Z defined by the set $\{(a, b) : a \equiv b \pmod{m}\}$ is an equivalence relation. This relation is called the **congruence modulo** m relation on the set of integers. The equivalence classes of this relation are called **congruence classes modulo** m. The congruence class of an integer x modulo m is dented by $[x]_m$.

0.68. Find the congruent classes modulo 5: **(a)** $[0]_5$, **(b)** $[1]_5$, and **(c)** $[2]_5$.

0.69. Prove that $\{[i]_m : i = 0, 1, 2, \ldots, (m-1)\}$ is a partition of the set of integers.

0.70. Let f be a function from A to A. Let R be the relation on A defined by $\{(x, y) : f(x) = f(y)\}$. Prove that R is an equivalence relation on A. What is the equivalence class?

0.71. Suppose that R is an equivalence relation on a nonempty set A. Show that there is a function f with A as domain such that (x, y) is in R if and only if $f(x) = f(y)$.

0.72. Suppose that $\{(a, b), (c, d)\}$ is in R whenever a, b, c, d are positive integers and $ad = bc$. Show that R is an equivalence relation on the set of positive integers.

0.73. Let $X = \{1, 2, 3, 4, 5, \ldots, 15\}$. Let R be the relation on X defined by $(x, y) \in R$ if $(x - y)$ is divisible by 3. Prove that R is an equivalence relation on X. Find the equivalence class.

0.74. Let R be a transitive and reflexive relation on a set A. If S is a relation on A such that (x, y) is in S if and only if both (x, y) and (y, x) are in R, prove that S is an equivalence relation on A.

0.75. Prove that a reflexive relation R on a set A is an equivalence relation on A if and only if (x, y) and (x, z) in R implies that (y, z) is in R.

0.76. If S is the relation defined on the set of real numbers as $S = \{(x, y) : x \leqslant y\}$, show that S is a partial order on the set of real numbers.

0.77. If S is the relation defined on the set of positive integers as $S = \{(x, y) : x \text{ divides } y\}$, show that S is a partial order on the set of positive integers.

0.78. Let $X = \{a, b, c\}$ and S be the partial order defined on the powerset $P(X)$ defined as $S = \{(A, B) : A \text{ is a subset of } B\}$. List the elements of S.

0.79. Draw the Hasse diagram of the partial order S of Problem 0.78.

0.80. Let $X = \{1, 2, 3, 4, 5, 6, 7, 8, 9\}$ and $S = \{(m, n), \text{ where } m \text{ divides } n\}$ be a partial order on X. Draw the Hasse diagram that represents S. Locate a chain in X.

0.81. Prove that if R is a partial order on the set A, its inverse R^{-1} also is a partial order on A. The partially ordered set (A, R^{-1}) is called the **dual** of the partially ordered set (A, R).

0.82. Let $A = \{2, 3, 4, 6, 8, 12, 16, 24\}$ and S be the partial order relation on A defined by $S = \{(a, b) : a \text{ divides } b\}$. Find **(a)** the minimal elements in A, **(b)** the maximal elements in A, and **(c)** the upper bounds of the set $B = \{4, 6, 12\}$.

0.83. Draw the Hasse diagram of the PO set in Problem 0.82.

0.84. Prove that (a) every finite partially ordered set has a maximal element and a minimal element, and (b) every finite linearly ordered set has a greatest element and a least element.

0.85. Prove by induction that $1^k + 2^k + 3^k + \cdots + n^k$ is equal to (a) $n(n + 1)(2n + 1)/6$ when $k = 2$, and (b) $[n(n + 1)/2]^2$ when $k = 3$.

0.86. Prove by induction that $1.2 + 2.3 + 3.4 + \cdots + n(n + 1)$ is equal to $n(n + 1)(n + 2)/3$.

0.87. Prove by induction that $1/1.2 + 1/2.3 + 1/3.4 + \cdots + 1/n(n + 1)$ is equal to $n/(n + 1)$.

0.88. Show that $n^3 + 2n$ is divisible by 3 for all positive integers n.

0.89. Prove that $1.2.3 + 2.3.4 + 3.4.5 + \cdots + n(n + 1)(n + 2)$ is equal to $n(n + 1)(n + 2)(n + 3)/4$.

0.90. Prove that $1^2/1.3 + 2^2/3.5 + \cdots + n^2/(2n - 1)(2n + 1)$ is equal to $[n(n + 1)]/2(2n + 1)$.

0.91. Show that the sum of the cubes of any three consecutive positive integers is divisible by 9.

0.92. Show that for any positive integer n greater than 1, the sum $1 + 1/\sqrt{2} + 1/\sqrt{3} + \cdots + 1/\sqrt{n}$ is greater than $\sqrt{n}$.

0.93. Prove: $(1 - 1/2)(1 - 1/3) \cdots (1 - 1/n) = 1/n$.

0.94. Prove: $2^n > n^2$ whenever $n > 4$.

0.95. Show that $1/(n + 1) + 1/(n + 2) + \cdots + 1/(2n)$ is greater than 13/24 whenever n is greater than 1.

0.96. Prove that $7^n - 1$ is divisible by 6.

0.97. Prove that $11^n - 6$ is divisible by 5.

0.98. Show that $6.7^n - 2.3^n$ is divisible by 4.

0.99. Prove: $3^n + 7^n - 2$ is divisible by 8.

0.100. Prove De Morgan's laws:
  (a) The absolute complement of the intersection of n subsets of a universal set is equal to the union of the absolute complements of these n sets.
  (b) The absolute complement of the union of these n sets is the intersection of their absolute complements.

0.101. Show that the cardinality of the powerset of a set with n elements is 2^n.

0.102. Prove that if S is a transitive relation on a set A, then S^n is a subset of S for $n = 1, 2, 3, \ldots$.

0.103. Suppose that f is recursively defined as $f(0) = 1$ and $f(n + 1) = 3f(n) + 5$. Find $f(1), f(2)$, and $f(3)$.

0.104. Give a recursive definition of x^n when x is a real number and n is a nonnegative integer.

0.105. Give a recursive definition of f where $f(n)$ is the sum of the first n positive integers.

0.106. Give a recursive definition of the set of **(a)** all integers, **(b)** all positive odd integers, **(c)** all negative even integers and **(d)** all even integers.

0.107. Construct the truth table of the statement $p \to (p \lor q)$ and determine whether it is a tautology or a contradiction or neither.

0.108. Show that $(p' \lor q) \land (p \land (p \land q))$ is equivalent to $(p \land q)$.

0.109. Examine whether $(p \to q) \to r$ is a tautology or a contradiction.

0.110. Construct the truth table of the statement $[(p \to q) \land (q \to p)] \leftrightarrow (p \leftrightarrow q)$ and determine whether it is a contradiction.

0.111. Construct the truth table of $q \leftrightarrow (p' \lor q')$.

0.112. Suppose that p and r are false statements and q and s are true statements. Find the truth values of
    (a) $(p \to q) \to r$
    (b) $(s \to (p \land r')) \land ((p \to (r \lor q)) \land s)$

0.113. Find the truth assignments of $p, q, r, s,$ and t such that the following are satisfiable:
    (a) $(p \land q \land r) \to (s \lor t)$
    (b) $(p' \land q') \lor r'$

0.114. Show that $(p \to q) \to (p' \lor q)$ is a tautology.

0.115. Prove: $(p \land q) \land (p \lor q)'$ is a contradiction.

0.116. Show that $((p \to q) \to q) ((p - q)' \lor q)$ is a tautology.

CHAPTER

1

Combinatorics

1.1 TWO BASIC COUNTING RULES

Combinatorics is one of the fastest-growing areas of modern mathematics. It has many applications to several areas of mathematics and is concerned primarily with the study of finite or discrete sets (much as the set of integers) and various structures on these sets, such as arrangements, combinations, assignments, and configurations. Broadly speaking, three kinds of problems arise while studying these sets and structures on them: (1) the **existence problem,** (2) the **counting problem,** and (3) the **optimization problem.** The existence problem is concerned with the following question: Does there exist at least one arrangement of a given kind? The counting problem, on the other hand, seeks to find the number of possible arrangements or configurations of a certain pattern. The problem of finding the most efficient arrangement of a given pattern is the optimization problem. In this chapter we study techniques for solving problems that involve counting. These techniques form a basis for the study of **enumerative combi-**

natorics, which is really the theory of counting where results involving counting are obtained without carrying out the exact counting process, which could be tedious.

Suppose that there are 10 mathematics majors and 15 computer science majors in a class of 25 and we are required to choose a student from the class to represent mathematics *and* another student to represent computer science. Now there are 10 ways of choosing a mathematics major and 15 ways of choosing a computer science major from the class. Furthermore, the act of choosing a student from one area in no way depends on the act of choosing a student from the other. So it is intuitively obvious that there are $10 \times 15 = 150$ ways of selecting a representative from mathematics and a representative from computer science. On the other hand, if we are required to select one representative from mathematics *or* from computer science, we have only $10 + 15 = 25$ ways of accomplishing this. In the former case we used the multiplication rule of counting and in the latter the addition rule. These two rules can be stated formally as follows.

MULTIPLICATION RULE (The Rule of Sequential Counting)

Suppose that there is a sequence of r events $E_1, E_2, \ldots, E_r$ such that (1) there are n_i ways in which E_i ($i = 1, 2, \ldots, r$) can occur, and (2) the number of ways an event in the sequence can occur does not depend on how the events in the sequence prior to that event occurred. Then there are $(n_1) \cdot (n_2) \cdot \ldots \cdot (n_r)$ ways in which all the events in the sequence can occur.

ADDITION RULE (The Rule of Disjunctive Counting)

Suppose that there are r events $E_1, E_2, \ldots, E_r$ such that (1) there are n_i outcomes for E_i ($i = 1, 2, \ldots, r$), and (2) no two events can occur simultaneously. Then there are $(n_1) + (n_2) + \cdots + (n_r)$ ways in which one of these r events can take place.

These two elementary rules are very useful in solving counting problems without carrying out explicit enumeration. However, if one is not careful, they are likely to be misused, producing erroneous results, as may be seen from some of the examples we discuss in what follows.

Example 1.1.1

There are five characters—two letters of the alphabet followed by three digits—which appear on the back of one series of a microcomputer made by an electronics company. The number of possible computers manufactured in this series is (1) $26 \times 26 \times 10 \times 10 \times 10 = 676,000$ if characters can repeat, (2) $26 \times 25 \times 10 \times 10 \times 10 = 650,000$ if letters cannot repeat, and (3) $26 \times 25 \times 10 \times 9 \times 8 = 468,000$ if no characters can repeat. We use the multiplication rule here.

Example 1.1.2

A professor has 25 students in her advanced calculus course and 31 students in her statistics course. Thirteen students have signed up for both the courses. There are three events here, no two of which can occur simultaneously: (1) The event that a student chosen at random has signed up for advanced calculus but not for statistics, and this can happen in 12 ways; (2) the event that a student chosen at random has signed up for statistics but not for advanced calculus, and this can happen in 18 ways; and (3) the event that a student chosen at random has signed up for both the courses and this can happen in 13 ways. By the addition rule one of these events can occur in $12 + 18 + 13 = 43$ ways. In other words, the professor has 43 students in both the courses together. Notice that the event that a student chosen at random takes advanced calculus and the event that a student chosen at random takes statistics can occur simultaneously. So we cannot apply the addition rule to the two events to conclude that the professor has a total of $25 + 31 = 56$ students.

Example 1.1.3

In a sightseeing group there are 8 Austrians, 5 Brazilians, and 6 Canadians. So by the multiplication rule there are 40 ways of choosing an Austrian and a Brazilian, 48 ways of choosing an Austrian and a Canadian, and 30 ways of choosing a Brazilian and a Canadian. Next, by the addition principle, there are $40 + 48 + 30 = 118$ ways of selecting a pair of individuals of distinct nationalities from this group of tourists. A team of 3 tourists of distinct nationalities can be chosen in $8 \times 5 \times 6$ ways, whereas a typical representative can be chosen in $8 + 5 + 6$ ways.

Example 1.1.4

The number of odd integers between 0 and 99 is obviously 50. We may invoke the multiplication rule to get this result. Any integer between 0 and 99 has a unit digit and a tens digit if we write 0, 1, 2, . . . , 9 as 00, 01, 02, . . . , 09. Let E be the event of choosing a digit for the unit digit. This can be done in 5 ways. Next, let F be the event of choosing a digit for the tens digit. This can be done in 10 ways. Notice that the number of ways that E can occur does not depend on how F can occur, and vice versa. So the sequence E, F (or for that matter, the sequence F, E) can occur in 50 ways.

Example 1.1.5

Suppose that we are interested in finding the number of odd integers between 0 and 100 with *distinct* digits. Let E and F be as in Example 1.1.4. E can be done in 5 ways as before. After that F can occur in 9 ways. The number of ways that F can occur does not depend on how E occurs. So by the multiplication rule the sequence E, F can occur in 45 ways, and consequently, there are 45 such integers. On the other hand, if F is the first event, it can occur in 10 ways. Subsequently, the second event E can be done in 5 ways if the tens digit is even, and in 4 ways if the tens digit is odd. In other words, the number of ways in which E occurs depends on how the event F occurs. So we cannot apply the multiplication rule to the sequence F, E in this case.

Example 1.1.6

Suppose that X is a set with n elements. List the elements of X as 1, 2, ... , n and consider the following sequence of n events: The first event is to decide whether or not to pick the first element, the second event is to decide whether or not to pick the second elment, and so on. Each event can occur in 2 ways and the number of ways that any of these events in the sequence can occur does not depend on how the previous events in the sequence occurred. Thus any set with n elements has 2^n subsets, by the multiplication rule. The class of all subsets of the set X is the power set of X and is denoted by $P(X)$ as mentioned in Chapter 0.

1.2 PERMUTATIONS

Consider a collection X of n *distinct* objects. An *r*-**permutation of** X is an arrangement in a row of any r objects from X. Of course, r is at most n. Thus if X is the collection of the first 5 letters a, b, c, d, and e, then edcb, dbea, and bdca are some of the several 4-permutations of X. The total number of r-permutations of a collection of n distinct objects is denoted by $P(n, r)$. Any r-permutation here can be considered as a sequence of r events in which the number of ways an event can occur does not depend on how the events prior to that event occur. So we use the multiplication rule of counting to conclude that $P(n, r)$ is equal to $n(n - 1)(n - 2) \cdots (n - r + 1)$ since any arbitrary object from X can be chosen in n ways and having chosen that, a second arbitrary object can be chosen in $(n - 1)$ ways, and so on, until all r objects are chosen.

Permutations and the Allocation Problem

We can approach this process of making arrangements of objects from a different point of view. Consider a set of n *distinct* locations arranged in a definite order and we are required to allocate r *distinct* objects to these locations such that no location can receive more than one object. Then the number of ways of allocating these r objects to the n locations is also $P(n, r)$ by the multiplication rule since any arbitrary object can be sent to one of the locations in n ways, and subsequently another one can be sent in $(n - 1)$ ways, and so on.

Example 1.2.1

If $X = \{1, 2, 3, 4, 5, 6, 7\}$ and $r = 3$, the number of r-permutations of X is $7 \times 6 \times 5 = 210$.

Any n-permutation of a set X with n elements is simply called a **permutation of** X and the number $P(n, n)$ of permutations of X is $n(n - 1)(n - 2) \cdot \cdot \cdot \cdot \cdot 3 \cdot 2 \cdot 1$, which is denoted by the factorial function $n!$. It is easy to see that $P(n, r) = n!/(n - r)!$. (We define $0! = 1$.)

[The positive integer $n!$ can be extremely large even when n is a small two-digit number. It is more than 3.6 million when $n = 10$ and it is approximately equal to $(2.433)(10^{18})$ when $n = 20$.]

Circular and Ring Permutations

Example 1.2.2

Consider a collection of 5 stones of different colors: blue (B), green (G), red (R), pink (P), and white (W).

(a) The number of ways of making a tiepin on which these 5 stones are to be placed horizontally is, of course, 5!.

(b) In how many ways can we make a tiepin on which these stones are placed in a circular pattern? The answer has to be less than 5! because some of the permutations considered in (a) are now not distinct. For example, if we rotate the permutation BGRPW once in the clockwise direction, we get the permutation GRPWB, and these two permutations are not distinct in a circular arrangement. If we fix one of the colors and then consider the permutations formed by the remaining 4 colors, these permutations are all distinct. For example, if we fix B and consider RGPW and RGWP, we get two permutations, BRGPW and BRGWP, which are distinct. Thus there are only (4!) such circular permutations.

(c) In how ways can we make a ring in which these stones are mounted? In a ring, there is no difference between a permutation and its "mirror image." For example, BGRPW and BWPRG are the same. For every permutation in (b), there is a mirror image. So the answer now is (4!)/2.

Thus the number of circular permutations of a set of n elements is $(n - 1)!$ and the number of ring permutations is $((n - 1)!)/2$.

Generalized Permutations

Let us now consider a collection X of n objects (*not necessarily distinct*) belonging to k different nonempty groups such that (1) all the objects in a group are identical, and (2) an object in a group is not identical to an object in another group. (For example, the letters in the collection a, b, a, b, b, d, e, e, d can be formed into four groups: one for a, one for b, one for d, and one for e.) Assume that there are n_i objects in group i where $i = 1, 2, \ldots, k$. Any arrangement in a row of these n objects is called a **generalized permutation of** X. (For example, LIN-ISOIL is a generalized permutation of the letters that appear in the word ILLI-NOIS.) The number of such generalized permutations is denoted by $P(n; n_1, n_2, \ldots, n_k)$, which will be $n!$ if all the objects in X are distinct.

THEOREM 1.2.1

If the collection X of n objects consists of k distinct nonempty groups such that group i has n_i identical objects (where $i = 1, 2, \ldots, k$), then the number of generalized permutations of X is $(n!)/(n_1!)(n_2!) \cdots (n_k!)$.

Proof:

If the objects belonging to group i were all distinct, there would have been $n_i!$ permutations for the elements in this group. So each generalized permutation gives rise to $N = (n_1!)(n_2!) \cdots (n_k!)$ permutations of X if X had distinct objects. If t is the total number of generalized permutations, we have $(t)(N) = n!$, from which the conclusion of the theorem follows. [Observe that if $k = n$, each group has exactly one element that is equivalent to the statement that the objects in X are distinct verifying that $P(n; 1, 1, \ldots, 1)$, where 1 is repeated n times, is equal to $n!$, as it should.] ◇

Example 1.2.3

The 9 letters that appear in the word CONSENSUS can be grouped into 6 groups: the group consisting of three S's, the group consisting of two N's, and four groups consisting of each of the four remaining distinct letters. The total number of generalized permutations in this case is $(9!)/(3!)(2!)(1!)(1!)(1!)(1!) = 30,240$.

If the total number of objects in any $(k - 1)$ of these k groups is r (where $r \leq n$), the formula for the number of generalized permutations can be expressed as $P(n, r)/(n_1!)(n_2!) \cdots (n_{k-1}!)$ since $n! = P(n, r) \cdot (n - r)!$ and $n_k = (n - r)$. For example,

$$P(15; 2, 3, 4, 6) = \frac{(15!)}{(2!)(3!)(4!)(6!)} = \frac{P(15, 9)}{(2!)(3!)(4!)} = \frac{P(15, 12)}{(2!)(4!)(6!)}$$

and so on. Thus if n_i ($i = 1, 2, \ldots, k$) are k positive integers whose sum is r where $r \leq n$ and if we define

$$P(n; n_1, n_2, \ldots, n_k) = \frac{P(n, r)}{(n_1!)(n_2!) \cdots (n_k!)}$$

we see that

    (1) $P(n; n_1, n_2, \ldots, n_{k-1}) = P(n; n_1, n_2, \ldots, n_{k-1}, m)$
where $m = n - (n_1 + n_2 + \cdots + n_{k-1})$
    (2) $P(n; r) = P(n; n - r) = P(n; r, n - r)$
    (3) $(r!) P(n; r) = P(n, r)$

Example 1.2.4

$$P(15; 2, 4, 4) = P(15; 2, 4, 4, 5) = \frac{(15!)}{(2!)(4!)(4!)(5!)} = \frac{P(15, 10)}{(2!)(4!)(4!)}$$

We now have the following generalization of Theorem 1.2.1, the proof of which is left as a simple exercise.

THEOREM 1.2.2

If there are n_i identical objects in group i ($i = 1, 2, \ldots, k$) and if r is the total number of the objects in these k groups, these r objects can be placed in n distinct locations, so that each location receives at most one object, in t ways, where $t = P(n; n_1, n_2, \ldots, n_k)$. In particular, if each group has exactly

one object, then $t = P(n, r)$, which is the number of r-permutations of a set
with n elements.   ◇

Example 1.2.5

Suppose that there are 100 spots (marked serially from 100 to 199) in the
showroom of a car dealership for displaying new cars in which 15 sports
cars, 25 compact cars, 30 station wagons, and 20 vans are to be parked
for display. Assume that the automobiles in each category are brand new
and identical in all respects, including color. The dealer can then park the
collection of 90 vehicles for display (leaving 10 blank spots in the lot) in
$P(100, 90)/(15!)(25!)(30!)(20!)$ ways.

1.3 COMBINATIONS

As in section 1.2, let X be a collection of n *distinct objects*. Any collection of
r distinct objects from X is called an **r-combination of** X. In other words, if X
is a set with n elements, any subset of X with r elements is an r-combination of
X. In an r-combination the order in which the r elements are chosen is not
important, unlike in the case of an r-permutation. The number of r-combinations
of a set with n elements is denoted by $C(n, r)$, which is precisely the number
of subsets of cardinality r. Thus there are $P(n, 2)$ *ordered pairs* and $C(n, 2)$
unordered pairs of two elements in a set of n elements. Of course, $C(n, 0) =$
$C(n, n) = 1$.

What is the relation between $C(n, r)$ and $P(n, r)$? Consider any subset A
of X with r elements. These r distinct elements can be arranged in $(r!)$ ways.
Thus there are $(r!)$ permutations associated with every r-element subset of X.
Of course, by definition, the number of r-element subsets of X is $C(n, r)$. Thus
the total number of r-permutations is the product of $(r!)$ and $C(n, r)$ by the
multiplication rule. So we have the following important theorem.

THEOREM 1.3.1

$$C(n, r) \cdot (r!) = P(n, r)$$

COROLLARY

$$C(n, r) = C(n, n - r)$$

Proof:

$$C(n, r) = \frac{(n!)}{r! (n - r)!} = \frac{(n!)}{(n - r)! r!}$$

$$C(n, n - r) = \frac{(n!)}{(n - r)! (n - (n - r))!} = \frac{(n!)}{(n - r)! \, r!}$$

In other words, if X is a set with n elements, the number of subsets of X with r elements is equal to the number of subsets of X with $(n - r)$ elements.

◇

Combinations and the Allocation Problem

As in the case of permutations, we can interpret combinations from a different point of view, as a problem of allocations. As before, let X be a set of n *distinct* locations arranged in a definite order and consider a collection of r objects that are *identical*. These objects are to be to allocated to these n locations such that no location receives more than one subject. Let t be the total number of ways of allocating these r objects. If all the objects were distinct, each *such* allocation would give rise to $(r!)$ allocations. In that case the total number of allocations would have been $(t)(r!)$. But the total number of allocations if the objects were distinct is $P(n, r)$. Thus $t = P(n, r)/(r!) = C(n, r)$.

THEOREM 1.3.2   (Pascal's Formula)

$$C(n, r) = C(n - 1, r) + C(n - 1, r - 1)$$

Proof:

Let X be a set with n elements and Y be any subset of X with $(n - 1)$ elements. Let t be the element of X that is not in Y. Every r-element subset of X is either a r-element subset of Y or the union of a subset of Y with $(r - 1)$ elements and the singleton set consisting of t. In the former category there are $C(n - 1, r)$ sets and in the latter there are $C(n - 1, r - 1)$ sets. In other words, the total number of subsets of X with r elements is the sum of $C(n - 1, r)$ and $C(n - 1, r - 1)$.

◇

Pascal's formula is an example of a *combinatorial identity*, which was proved using a *combinatorial argument*. This identity can be proved algebraically also. A few combinatorial identities are given at the end of this chapter as exercises. Here is another example.

Example 1.3.1

$$C(2n, 2) = 2C(n, 2) + n^2$$

Proof:

Let X be any set with $2n$ elements that is partitioned into two sets Y and Z, each containing n elements. The number of subsets of X with two elements is $C(2n, 2)$. Any subset of X has two elements if and only if it belongs to one of the following three classes: (1) the class of all subsets of Y with two elements, (2) the class of all subsets of Z with two elements, and (3) the class of all subsets of X with two elements such that each subset in this class has one element from Y and one element from Z. Classes (1) and (2) have $C(n, 2)$ sets each. An element from Y can be chosen n ways and an element from Z can be chosen in n ways. So class (3) has $(n)(n)$ sets. Thus the number of subsets of X with two elements is $C(n, 2) + C(n, 2) + (n)(n)$.

The Allocation Problem and Generalized Combinations

Now consider a collection of n objects (not necessarily distinct) belonging to k distinct groups, as in the hypothesis of Theorem 1.2.1. The n_1 identical objects of group 1 can be placed in the set of n locations (such that no location receives more than one object) in $C(n, n_1)$ ways. Then the n_2 objects of the next group can be placed in $C(n - n_1, n_2)$ ways. We proceed in this way until all spots are filled. By the multiplication rule the total number of ways in which all the n spots can be filled is

$$C(n, n_1) \cdot C(n - n_1, n_2) \cdot C(n - n_1 - n_2, n_3)$$

$$\cdots \cdot C(n - n_1 - n_2 - \cdots - n_{k-1}, n_k)$$

and this number is denoted by $C(n; n_1, n_2, \ldots, n_k)$.

There is another way of looking at this allocation process. Suppose that X is a collection of n *distinct* objects and these n objects are to be allocated to k locations so that location i gets n_i objects ($i = 1, 2, \ldots, k$) where $n_1 + n_2 + \cdots + n_k = n$. Then any n_1 objects can be selected from X and allocated to location 1 in $C(n, n_1)$ ways. Next, from the remaining $(n - n_1)$ objects in X, any n_2 objects can be allocated to location 2 in $C(n - n_1, n_2)$ ways. We proceed like this until all the objects are exhausted.

We have the following result connecting generalized permutations and combinations.

THEOREM 1.3.3

$$P(n; n_1, n_2, \ldots, n_k) = C(n; n_1, n_2, \ldots, n_k)$$

where $n_1 + n_2 + \cdots + n_k \leq n$.

Proof:

$$C(n, n_1) = \frac{(n!)}{(n_1)!\,(n - n_1)!}$$

$$C(n - n_1, n_2) = \frac{(n - n_1)!}{(n_2)!\,(n - n_1 - n_2)!}$$

$$C(n - n_1 - n_2, n_3) = \frac{(n - n_1 - n_2)!}{(n_3)!\,(n - n_1 - n_2 - n_3)!}$$

$$\vdots$$

Multiplying out the terms on the right-hand side of the equation, we get $C(n; n_1, n_2, n_3, \ldots, n_k) = (n!)/(n_1)!(n_2)! \cdots (n_k)!$, and thus the theorem is established.

Observe that if all the objects in X are distinct and if we take r objects, we have $P(n; 1, 1, \ldots, 1) = C(n; 1, 1, \ldots, 1)$, where 1 is repeated r times. Now $P(n; 1, 1, \ldots, 1) = P(n, r)/(1!)(1!) \cdots (1!) = P(n, r)$ and $C(n; 1, 1, \ldots, 1) = C(n, 1) \cdot C(n - 1, 1) \cdots C(n - r + 1, 1) = n(n - 1) \cdots (n - r + 1)$ and we once again see that $P(n, r) = n(n - 1)(n - 2) \cdots (n - r + 1)$. ◇

COROLLARY

$$C(n; r) = C(n; n - r) = C(n; r, n - r) = C(n, r) = C(n, n - r)$$

Proof:

$$C(n; r) = C(n, r) \cdot C(n - r, n - r) = C(n, r)$$

$$C(n; n - r) = C(n, n - r) \cdot C(n - (n - r), r) = C(n, n - r) = C(n, r)$$

$$C(n; r, n - r) = C(n, r) \cdot C(n - r, n - r) = C(n, r)$$

[Notice that $C(n, r) = C(n; r) = P(n; r)$, but $P(n, r) = (r!)P(n; r)$.] ◇

Example 1.3.2

$$P(15; 3, 5, 7) = \frac{(15!)}{(3!)(5!)(7!)} = \frac{P(15,8)}{(3!)(5!)} = P(15; 3, 5)$$

$$C(15; 3, 5, 7) = C(15, 3)C(12, 5)C(7, 7)$$

$$= C(15, 3)C(12, 5) = C(15; 3, 5)$$

Also,

$$C(15, 3) = \frac{(15!)}{(3!)(12!)} \quad \text{and} \quad C(12, 5) = \frac{(12!)}{(5!)(7!)}$$

Thus

$$C(15; 3, 5, 7) = \frac{(15!)}{(3!)(5!)(7!)} = P(15; 3, 5, 7)$$

The Multinomial Theorem

THEOREM 1.3.4   (The Multinomial Theorem)

In a typical term in the expansion of $(x_1 + x_2 + \cdots + x_k)^n$ the variable x_i $(i = 1, 2, \ldots, k)$ appears n_i times (where $n_1 + n_2 + \cdots + n_k = n$) and the coefficient of this typical term is $C(n; n_1, n_2, \ldots, n_k)$.

Proof:

The first part of the assertion is obvious since the expansion is the product of n expressions where each expression is the sum of the k variables. A typical term here is nothing but a generalized permutation of n objects in a collection X consisting of k groups, and therefore the coefficient of this typical term is the number of such generalized permutations.

$\diamondsuit$

Example 1.3.3

The coefficient of $a^3b^2c^6d^4$ in the expansion of $(a + b + c + d)^{15}$ is $(15!)/(3!)(2!)(6!)(4!)$.

Example 1.3.4   (The Binomial Theorem)

The multinomial theorem when $k = 2$ is known as the binomial theorem, which can be stated as

$$(x + y)^n = \Sigma \, C(n, n - r)x^{n-r}y^r \quad \text{where } r \text{ varies from 0 to } n$$

[The right-hand side of this equation is called the **binomial expansion** of $(x + y)^n$. The coefficients $C(n, r)$ that appear in the binomial expansion are called **binomial coefficients.**]

The binomial coefficients of $(x + y)^n$ can be computed if we know the binomial coefficients of $(x + y)^{n-1}$ by using Pascal's formula: $C(n, r) = C(n - 1, r) + C(n - 1, r - 1)$. So the binomial coefficients can be arranged in the form of a triangle known as **Pascal's triangle:**

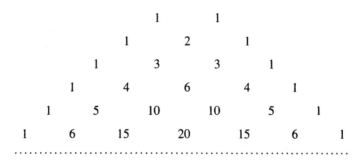

```
            1        1
        1        2        1
    1        3        3        1
1        4        6        4        1
1    5        10       10       5        1
1    6    15       20       15       6        1
```

In this representation, the $(n + 1)$ consecutive binomial coefficients of the binomial expansion of $(x + y)^n$ appear in the nth row. Notice that a typical element in a row (other than the first and the last) is the sum of the two terms just above that element which appear in the preceding row—and that is exactly the content of Pascal's formula. In each row the first element as well as the last element is 1, indicating the fact that if A is a set with n elements, then there is only one subset of A with n elements and there is only subset of A with no elements.

Partitioning of a Finite Set

Given a set A of cardinality n, a combinatorial problem of interest is to find the number of ways A can be partitioned into k subsets such that subset A_i ($i = 1$, $2, \ldots, k$) has exactly n_i elements. For example, if $A = \{1, 2, 3, 4, 5, 6\}$, the problem will be to find the number of ways of partitioning A into (1) two subsets such that one has 2 elements and the other has 4 elements or (2) two subsets such that each has 3 elements or (3) three subsets such that each has 2 elements, and so on. This problem is equivalent to the allocation problem of allocating n distinct objects to k locations discussed earlier when the cardinalities of each set in the partition are distinct as in (1). There are $C(6; 2, 4)$ ways of allocating the 6 elements from A to two locations so that location 1 gets 2 elements and location 2 gets 4 elements. The number of ways of partitioning A into two sets so that one set has 2 elements and the other has 4 is also $C(6; 2, 4)$. But when the subsets in a partition have equal cardinalities, we have to take care of those situations where repetitions occur. For example, if $P = \{1, 2, 3\}$ and $Q = \{4, 5, 6\}$, the partition $\{P, Q\}$ and the partition $\{Q, P\}$ are the same. But allocating P to location 1 and Q to location 2 is not the same as allocating Q to location

1 and P to location 2. The number of partitions of A into two subsets of equal cardinality is $C(6; 3, 3)/2$. More generally, we have the following result, which is an extension of the allocation theorem and the multiplication rule.

THEOREM 1.3.5

The number of ways of partitioning a set of cardinality n into a class consisting of p_i subsets each of cardinality n_i $(i = 1, 2, \ldots, k)$ where no two of the numbers n_i are equal is

$$\frac{(n!)}{(p_1!)(n_1!)^{p_1}(p_2!)(n_2!)^{p_2} \cdots (p_k!)(n_k!)^{p_k}}$$

Example 1.3.5

(a) The number of ways of *allocating* 43 students into 7 *different* dormitories such that the first two get 5 students each, the next three get 6 students each, the sixth dormitory gets 7 students, and the seventh dormitory gets 8 students is

$$\frac{(43!)}{(5!)(5!)(6!)(6!)(6!)(7!)(8!)}$$

(b) The number of ways of *dividing* 43 students *into 7 groups* such that there are 5 students in each of 2 groups, 6 students in each of 3 groups, 7 students in one group, and 8 students in one group is

$$\frac{(43!)}{(2!)(5!)(5!) \cdot (3!)(6!)(6!)(6!) \cdot (7!) \cdot (8!)}$$

1.4 MORE ON PERMUTATIONS AND COMBINATIONS

If X is a set with n elements, we know that an r-permutation of X is an arrangement of elements from X in which no elements repeat. Similarly, an r-combination is a selection of elements from X in which no elements repeat. In both cases r cannot exceed n. If we allow repetitions, there is no restriction on r. (Since X is a set, the n elements in it are all distinct.) An r-**sequence of** X is an arrangement of r elements from X in which the elements may repeat, but the order in which these elements appear is important. For example, *aabdac* and *aadbac* are two distinct 6-sequences from the set $X = \{a, b, c, d\}$. Any r-permutation is obviously an r-sequence. On the other hand, any r-sequence with distinct elements is an r-permutation. A simple application of the multiplication rule shows that the number of r-sequences in a set with n elements is n^r.

Any collection of r objects (not necessarily distinct) chosen from a set X of n elements is called an r-**collection** from X. Unlike an r-selection, the order in which the elements are chosen is not important in an r-collection. The 4-collection $[a, a, b, c]$ is not different from the 4-collection $[a, b, c, a]$. Both represent the same 4-collection. Any r-combination is an r-collection. If the elements in an r-collection are distinct, then it is an r-combination. For example, if $X = \{a, b, c, d\}$ and if $r = 3$, the set of all 3-selections from X will include every subset of X with 3 elements and selections such as $\{a, a, a\}$, $\{a, b, b\}$, $\{d, a, d\}$ and so on. On the other hand, if $r = 5$, the collection $\{a, b, b, b, d\}$ is one of the ways of choosing 5 elements from X, and no subset of X can be a 5-collection.

Given a set of cardinality n and an arbitrary positive integer r, in how many ways can one choose r elements (with repetitions) from X? Here is the answer.

THEOREM 1.4.1

If X is a set of cardinality n, then the number of r-collections from X is $C(r + n - 1, n - 1)$, where r is any positive integer.

Proof:

Let $X = \{1, 2, 3, \ldots, n\}$. Let u be an r-collection from X in which 1 repeats x_1 times, 2 repeats x_2 times, $\ldots$, and n repeats x_n times. This r-collection can be represented as

$$u = [\underbrace{1 \cdots 1}\quad \underbrace{2 \cdots 2}\quad \underbrace{3 \cdots 3}\qquad\qquad \underbrace{n \cdots n}]$$

where the notation $i \cdots i$ means that the symbol i repeats x_i times. Similarly, let v be another r-collection in which 1 repeats y_1 times, 2 repeats y_2 times, $\ldots$, and n repeats y_n times. Then

$$v = [\underbrace{1 \cdots 1}\quad \underbrace{2 \cdots 2}\quad \underbrace{3 \cdots 3}\qquad\qquad \underbrace{n \cdots n}]$$

where the notation $i \cdots i$ means that the symbol i repeats y_2 times. Observe that in the representation of u as well as in the representation of v, there is a gap between 1 and 2, a gap between 2 and 3, $\ldots$, a gap between $(n - 1)$ and n. In each representation there are $(n - 1)$ gaps. What distinguishes one r-collection from another is where these blank spots are located in a typical representation.

Each representation has r symbols and $(n - 1)$ gaps. So each representation can be considered as set of $r + n - 1$ distinct locations. All the $n - 1$ blanks are identical. An allocation of these $(n - 1)$ blanks to the $(r + n - 1)$

locations defines an r-collection. Thus the number of distinct r-collections is the same as the number of ways of allocating $(n - 1)$ identical objects to $(r + n - 1)$ distinct locations so that each location receives at most one object. This number is $C(r + n - 1, n - 1)$, as we saw in Section 1.3. $\diamond$

Example 1.4.1

Let $X = \{a, b, c, d\}$. The total number of 5-selections from X will be $C(5 + 4 - 1, 4 - 1) = 56$.

The following theorem is an equivalent version of Theorem 1.4.1.

THEOREM 1.4.2

(a) The number of distinct solutions in nonnegative integers of the linear equation (in n variables) $x_1 + x_2 + \cdots + x_n = r$ is $C(r + n - 1, n - 1)$.

(b) The number of distinct solutions in nonnegative integers of the linear inequality (in n variables) $x_1 + x_2 + \cdots + x_n \leq r$ is $C(r + n, n)$.

(c) The number of terms in the multinomial expansion of $(x_1 + x_2 + \cdots + x_n)^r$ is $C(r + n - 1, n - 1)$.

Proof:

(a) Every solution $x_i = s_i$ $(i = 1, 2, \ldots, n)$ in nonnegative integers corresponds to a collection of r elements (from the set X consisting of the n variables) in which x_i repeats s_i times, where $s_i \leq r$, and vice versa. The number of such collections is $C(r + n - 1, n - 1)$ by Theorem 1.4.1.

(b) Let y be a nonnegative variable such that $x_1 + x_2 + \cdots + x_n + y = r$. ($y$ is called the **slack variable**.) We now have a linear equation in $(n + 1)$ variables. A solution in nonnegative integers of this equation in $(n + 1)$ variables is a solution in nonnegative integers of the inequality in n variables, and vice versa. Thus the required number is $C(r + n, n)$.

(c) Each term in the expansion can be considered as a product of the n variables in which the sum of the exponents of the variables is r. Therefore, the number of terms in the expansion is equal to the number of collections of r elements from the set X consisting of the n variables where repetitions are allowed. $\diamond$

Example 1.4.2

In an undergraduate dormitory there are several freshmen, sophomores, juniors, and seniors.

(a) In how many ways can a team of 10 students be chosen to represent the dormitory?

(b) In how many ways can a team of 10 be chosen such that it has at least one freshman, at least one sophomore, at least two juniors, and at least two seniors?

Solution

(a) If p, q, r, and s are the number of students of each class in the team, then the number of ways the team can be chosen is equal to the number of solutions in nonnegative integers of the equation $p + q + r + s = 10$. So the answer is $C(13, 3) = 286$.

(b) In this case $p > 0$, $q > 0$, $r > 1$, and $s > 1$. Write $p = p' + 1$, $q = q' + 1$, $r = r' + 2$, and $s = s' + 2$. So the number of ways will be equal to the number of solutions in nonnegative integers of the equation $p' + q' + r' + s' + 6 = 10$ and the answer is $C(7, 3) = 35$.

The Allocation Problem in the General Setting

We now consider the problem of allocating r *identical* objects to n distinct locations such that each location can accommodate as many objects as necessary. In how many ways can we accomplish this? If the number of objects placed in location i is x_i (where $i = 1, 2, \ldots, n$), any solution of the equation $x_1 + x_2 + \cdots + x_n = r$ in nonnegative integers corresponds to a way of allocating these r objects to the n locations, and vice versa. Thus there are $C(r + n - 1, n - 1)$ ways of placing r identical objects in n distinct locations. We combine this observation with the Theorems 1.4.1 and 1.4.2 to make the following assertion:

THEOREM 1.4.3

Let

$L = $ the number of ways of choosing r elements (with repetitions) from a set that has n elements

$M = $ the number of ways of allocating r identical objects to n distinct locations

$N = $ the number of solutions in nonnegative integers of the equation $x_1 + x_2 + \cdots + x_n = r$

Then

$$L = M = N = C(r + n - 1, n - 1) \qquad \Diamond$$

We now summarize the four cases of permutations and combinations (without or with repetitions) of r elements from a set of n distinct elements and interpret these results as two models of counting as follows.

The selection model

The number of ways of selecting r elements from a set of n elements is:

1. $P(n, r)$ if the elements selected are distinct and the order in which they are selected is important.
2. $C(n, r)$ if the elements selected are distinct and the order in which they are selected is not important.
3. n^r if the elements selected are not necessarily distinct and the order is important.
4. $C(r + n - 1, n - 1)$ if the elements selected are not necessarily distinct and the order is not important.

The allocation model

The number of ways of allocating r objects to n distinct locations is:

1. $P(n, r)$ if the objects are distinct and no location can take more than one object.
2. $C(n, r)$ if the objects are identical and no location can take more than one object.
3. n^r if the objects are distinct and there is no restriction on the number of objects in a location.
4. $C(r + n - 1, n - 1)$ if the objects are identical and there is no restriction on the number of objects in a location.

These conclusions can be summarized in Table 1.4.1.

We conclude this section with the following theorem, which summarizes the various cases of allocations considered so far.

THEOREM 1.4.4

(a) If r is at most n, a collection of r *distinct* objects can be allocated to n locations, so that no location can receive more than one object in $P(n, r)$ ways.

(b) A collection of r *distinct* objects can be allocated to n locations in n^r ways if there is no restriction on the number of objects that a location can receive.

(c) If r is at most n, a collection of r *identical* objects can be allocated to n locations so that no location can receive more than one object in $C(n, r)$ ways.

(d) A collection of r *identical* objects can be allocated to n locations such that location i gets at least p_i objects in $C(r - p + n - 1, n - 1)$ ways, where $p = p_1 + p_2 + \cdots + p_n$. (Theorem 1.4.3 is a special case when each p_i is 0.)

TABLE 1.4.1

	Selection model	Allocation model
	(*X* is a set with *n* elements. Select *r* elements from *X*.)	(*X* is a set of *n* distinct locations. Allocate a collection of *r* objects to these locations.)
$P(n, r)$	Number of ways of arranging *r* elements (selecting where order is important) from *X* such that elements selected do not repeat	Number of ways of allocating *r* distinct objects to the *n* locations such that no location can receive more than one object
$C(n, r)$	Number of ways of choosing *r* elements (selecting where order is not important) from *X* such that elements selected do not repeat	Number of ways of allocating *r* identical objects to the *n* locations so that no location can receive more than one object
n^r	Number of ways of arranging *r* elements from *X* such that elements selected may repeat (no restriction on *r* and order is important)	Number of ways of allocating *r* distinct objects to the *n* locations if a location can receive more than one object
$C(p, q)$, where $p = r + n - 1, q = n - 1$	Number of ways of selecting *r* elements from *X* such that elements may repeat (no restriction on *r* and order is not important)	Number of ways of allocating *r* identical objects to the *n* locations if a location can receive more than one object

(e) Suppose that there are k types of objects such that type i has n_i objects $(i = 1, 2, \ldots, k)$. Objects belonging to the same type are identical and two objects belonging to two different types are not identical. Then these $n_1 + n_2 + \cdots + n_k$ objects can be allocated to n locations so that no location can receive more than one object in $P(n; n_1, n_2, \ldots, n_k)$ ways.

(f) A collection of $n_1 + n_2 + \cdots + n_k$ *distinct* objects can be allocated to k locations so that location i receives exactly n_i objects $(i = 1, 2, \ldots, k)$ in $C(n; n_1, n_2, \ldots, n_k)$ ways.

(g) $P(n; n_1, n_2, \ldots, n_k) = C(n; n_1, n_2, \ldots, n_k) = P(n, r)/[(n_1!)(n_2!) \cdot \cdots \cdot (n_k!)]$ where $n_1 + n_2 + \cdots + n_k = r$.

1.5 THE PIGEONHOLE PRINCIPLE

This is a principle that is very obvious and looks very simple, as though it has no major significance. However, in practice it is of great importance and power since its generalizations involve some profound and deep results in combinatorial theory and in number theory. We are using the pigeonhole principle when we say that in any group of three people at least two are of the same sex. Suppose that the newly formed computer science department in a college has 10 faculty members and only 9 offices to accommodate them. Then the underlying idea behind the obvious assertion that at least one office will have more than one occupant is again the pigeonhole principle. If there are 19 faculty members instead of 10, at least one office will have more than two occupants. Similarly, if there are at least 367 students in a residence hall, it is equally obvious that at least two of them will have the same birthday. It is reported that the scalp of a human being has at most 99,999 hairs. So in any city whose population exceeds 4 million there will be at least 41 people (a bald scalp has no hair) with the same number of hairs! We can cite several examples like this.

The basic idea that governs all these instances is the simple fact known as the **Dirichlet pigeonhole principle,** which is stated formally as follows: If $n + 1$ or more pigeons occupy n pigeonholes, there will be more than one pigeon in at least one pigeonhole. More generally, if $kn + 1$ or more pigeons occupy n pigeonholes, there will be more than k pigeons in at least one pigeonhole, where k is a positive integer.

Here are some examples to illustrate this principle.

Example 1.5.1

In a round-robin tournament (in which every player plays against every other player exactly once), suppose that each player wins at least once. Then there are at least two players with the same number of wins. Suppose that there are n players. The number of wins for a player is 1 or 2 or 3 . . . or $(n - 1)$. These $(n - 1)$ numbers correspond to $(n - 1)$ pigeonholes in which the n players are to be housed. So at least two of them should be in the same pigeonhole and they have the same number of wins.

Example 1.5.2

There are 18 residence halls in campus. The dean of students would like to conduct a survey in any one of these halls about the use of microcomputers, and to do this she has to form a committee of 5 students from the hall chosen for the survey. An advertisement in the campus paper asks for volunteers from these 18 halls. At least how many responses to the advertisement are sufficient before the dean can choose a hall and form a committee?

Solution.   The answer is $(4)(18) + 1 = 73$ by the pigeonhole principle.

Example 1.5.3

A bag contains exactly 5 red, 8 blue, 10 white, 12 green, and 7 yellow marbles. Find the least number of balls to be chosen which will guarantee that there will be (a) at least 4 marbles of the same color, (b) at least 6 marbles are of the same color, (c) at least 7 marbles are of the same color, and (d) at least 9 marbles are of the same color. (Here each color represents a pigeonhole. The number of pigeonholes is $n = 5$.)

Solution

(a) If at least 4 marbles are of the same color, there is a pigeonhole whose occupancy is more than 3. So by applying the generalized pigeonhole principle with $k = 3$, the number of marbles to be chosen is at least $(3) \cdot (5) + 1 = 16$.

(b) $n = 5$ and $k = 5$. So the number is 26.

(c) $n = 5$ and $k = 6$. Notice that there is an upper limit on the number of red marbles. There are only 5 red marbles. So in this case the required number is $[(6) \cdot (5) + 1] - (6 - 5) = 30$.

(d) Now $n = 5$ and $k = 8$ with upper bounds of 5 for red and 7 for yellow. So the number is $[(8) \cdot (5) + 1] - (8 - 5) - (8 - 7) - (8 - 8) = 37$.

If m and n are positive integers, then the **floor** of m/n is the largest integer that is less than or equal to m/n and the **ceiling** of m/n is the smallest integer greater than or equal to m/n. (For example, the floor of 38/9 is 4 and the ceiling is 5.)

The following extension of the pigeonhole principle is easily established.

THEOREM 1.5.1

(a) If m pigeons are allotted to n pigeonholes, then at least one hole has more than k pigeons, where k is the floor of $(m - 1)/n$.

(b) If $m = p_1 + p_2 + \cdots + p_n - n + 1$ pigeons (each p_i is a positive integer) are allotted to n pigeonholes, then the first pigeonhole has at least p_1 pigeons, or the second pigeonhole has at least p_2 pigeons, . . . , or the nth pigeonhole has at least p_n pigeons.

Proof:

(a) Now $(n) \cdot (k) \leq (m - 1) < m$. If the number of pigeons is exactly $n \cdot k$, it is possible to allocate k pigeons to each hole. But the number of pigeons is m, which is greater than $n \cdot k$. So there is at least one hole with more than k occupants.

(b) Here k = floor of $[(p_1 + p_2 + \cdots + p_n)/n] - 1$. So $(k + 1)$ is equal to or greater than at least one of the n integers.   ◇

Example 1.5.4

A bag contains exactly 6 red, 5 white, and 7 blue marbles. Find the least number of marbles to be selected which will ensure that either at least 3 red or at least 4 white or at least 5 blue marbles picked.

Solution

First Method (Using Theorem 1.5.1). Here $n = 3$, $p_1 = 3$, $p_2 = 4$, and $p_3 = 5$. So $m = (3 + 4 + 5) - 3 + 1 = 10$.

Second Method. Let the number of red, white, and blue marbles to be selected be x, y, and z, respectively. We require that x is at least 3 or y is at least 4 or z is at least 5. This situation will not happen if x is at most 2 and y is at most 3 and z is at most 4, which implies that $x + y + z$ is at most 9. Thus we have to select at least 10 marbles.

Example 1.5.5

In any group of 6 people there are 3 people known to one another or there are 3 total strangers.

Proof. Let $\{A, B, C, D, E, F\}$ be the set of 6 people and let Y be a room in which individuals known to A are seated. Let Z be the room in which individuals not known to A are seated. The five individuals B, C, D, E, and F have to be assigned to the two rooms Y and Z. So by the previous proposition either Y or Z has at least $k + 1$ individuals, where k = floor of $(5 - 1)/2 = 2$. See Figure 1.5.1. If there is a dotted line joining two names, these two individuals do not know each other. If there is a line joining two individuals, they know each other.

(a) Suppose that room Y has 3 or more people. Let B, C, and D be three individuals in Y. There are two possibilities: Either B, C, and D do not know one another, as in Figure 1.5.1(a), forming a group of 3 strangers, or at least 2 of them (say, C and D) know each other, as in Figure 1.5.1(b). In the latter case, these two individuals, C and D, along with A, form a group of 3 people who know each other.

(b) Suppose that room Z has 3 or more people. Let B, C, and D be 3 of the people in Z. There are two possibilities: Either these 3 individuals know one another as in Figure 1.5.1(c), forming a group of 3 individuals known to each other, or there are at least 2 individuals (say, C and D) who do not know each other. In the latter case, these 2 individuals, C and D, along with A, form a group of 3 strangers.

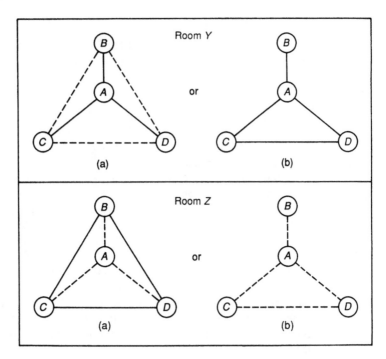

FIGURE 1.5.1

We conclude this brief exposition of the pigeonhole principle with two theorems due to Paul Erdös.

THEOREM 1.5.2

Let $X = \{1, 2, 3, \ldots, 2n\}$ and let S be any subset of X with $(n + 1)$ elements. Then there are at least two numbers in S such that one divides the other.

Proof:

Any number r in S can be represented as $r = 2^t \cdot s$, where t is a nonnegative integer and s is an odd number from X, called the *odd part* of r. There are at most n choices for s since there are n odd numbers in X. The n odd parts can be considered as n pigeonholes and the $(n + 1)$ numbers of S are to be allotted to these holes. In other words, there are two numbers x and y in S with the same odd part. Let $x = 2^t \cdot s$ and $y = 2^u \cdot s$. Then either x divides y, or vice versa.

THEOREM 1.5.3

Any sequence of $(n^2 + 1)$ distinct numbers contains a subsequence of at least $(n + 1)$ terms which is either an increasing sequence or a decreasing sequence.

Proof:

Let the sequence be a_i $(i = 1, 2, \ldots, n^2 + 1)$ and let t_i be the number of terms in the longest increasing subsequence that starts from a_i. If $t_i = n + 1$ for some i, we are done.

Suppose that $t_i \leq n$ every i. Let $H_j = \{a_i : t_i = j\}$, where $j = 1, 2, \ldots, n$. We thus have n pigeonholes $H_1, H_2, \ldots, H_n$ to which the $(n^2 + 1)$ numbers t_i are allotted. So by the generalized pigeonhole principle there is pigeonhole H_r containing more than k of these numbers where $k = $ floor of $[(n^2 + 1) - 1]/n = n$. So among the numbers t_i, at least $(n + 1)$ of them are equal.

We now establish that the $(n + 1)$ numbers in the sequence which correspond to these numbers in the pigeonhole H_r form a decreasing sequence. Let a_i and a_j be in H_r, where $i < j$. Either $a_i < a_j$ or $a_i > a_j$ since the elements in the sequence are all distinct. Suppose that $a_i < a_j$. Now $a_j \in H_r$ implies that there is a subsequence of length r starting from a_j. So $a_i < a_j$ implies that there is subsequence of length $(r + 1)$ starting from a_i. This is a contradiction, because there cannot be a subsequence of length $(r + 1)$ starting from a_i since a_i is an element of H_r. Thus $a_i > a_j$ whenever $i < j$. So any $(n + 1)$ elements in H_r will give rise to a strictly decreasing subsequence. $\diamondsuit$

Example 1.5.6

Illustrate Theorem 1.5.3 in the case of the sequences:
  (a) 15, 12, 5, 7, 9, 6, 3, 4, 10, 14
  (b) 15, 12, 9, 10, 7, 5, 4, 14, 3, 6

Solution

(a) Here $n = 3$, as there are 10 elements in the sequence and the corresponding t_i (10 of them) are 1, 2, 5, 4, 3, 3, 4, 3, 2, 1. Since t_3 is 5, there is an increasing subsequence of 5 elements starting from a_3, which is 5, 7, 9, 10, 14.

(b) Here the corresponding t_i are 1, 2, 3, 2, 2, 3, 2, 1, 2, 1. None of them exceeds 3. We get $H_1 = \{15, 14, 6\}$, $H_2 = \{12, 10, 7, 4, 3\}$, and $H_3 = \{9, 5\}$. The sequence coming out of the second set is a decreasing subsequence of the given sequence with 5 numbers.

1.6 *THE INCLUSION–EXCLUSION PRINCIPLE*

If X is any finite set, we denote by $N(X)$ the number of elements in X. Suppose that A and B are two finite sets with no elements in common. Then, obviously, $N(A \cup B) = N(A) + N(B)$. If, on the other hand, the intersection of A and B is nonempty, in order to compute the cardinality of $A \cup B$, we first find the sum $N(A) + N(B)$ as before. In this sum, the elements common to A and B are counted (included) twice—once while counting $N(A)$ and then while counting $N(B)$—so they have to be removed (excluded) once to obtain the total number of elements in their union.

For example, if there are 15 students in a class who take calculus, 12 students who take discrete mathematics, and 9 students who take both courses, then the number of students who take at least one of the two courses is $15 + 12 - 9 = 18$. See the Venn diagram in Figure 1.6.1 representing the universal set X of all students in the class, the set A of all students in the class who take calculus, and the set of B of all students who take discrete mathematics. The fact there are students in the class who take both the courses is made clear by showing that the set representing their intersection is the region common to A and B. The set to be excluded because it is included once with A and then with B is the subset $A \cap B$. Thus we can state the **inclusion–exclusion principle** involving two finite sets as follows: If A and B are two finite sets, then $N(A \cup B) = N(A) + N(B) - N(A \cap B)$.

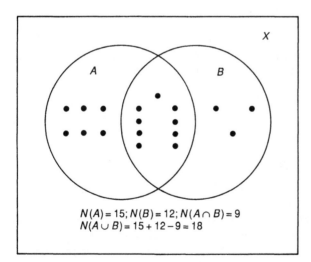

$N(A) = 15; N(B) = 12; N(A \cap B) = 9$
$N(A \cup B) = 15 + 12 - 9 = 18$

FIGURE 1.6.1

Example 1.6.1

Obtain the inclusion–exclusion rule involving three finite sets. Let A, B, and C be three finite sets and let $D = B \cup C$. Now $N(A \cup B \cup C) = N(A \cup D) = N(A) + N(D) - N(A \cap D)$ and $N(D) = N(B \cup C) = N(B) + N(C) - N(B \cap C)$. So

$$N(A \cup B \cup C) = N(A) + N(B) + N(C) - N(B \cap C) - N(A \cap D) \quad (*)$$

But $(A \cap D) = A \cap (B \cup C) = (A \cap B) \cup (A \cap C)$. So

$$N(A \cap D) = N(A \cap B) + N(A \cap C) - N((A \cap B) \cap (A \cap C))$$
$$= N(A \cap B) + N(A \cap C) - N(A \cap B \cap C)$$

On substituting for $N(A \cap D)$ in (*), we get

$$N(A \cup B \cup C) = [N(A) + N(B) + N(C)]$$
$$- [N(A \cap B) + N(A \cap C) + N(B \cap C)] + N(A \cap B \cap C)$$

which is the inclusion–exclusion rule for three sets.

Suppose that the sets we consider are all finite subsets of a certain finite set X with N elements. If A is a subset of X, the complement of A is denoted by A'. Then $A' = X - A$. So

$$N(A') = N - N(A) \qquad \text{for any subset } A \text{ contained in } X \qquad (i)$$

Next let A and B be two subsets of X. Then by (i), $N((A \cup B)') = N - N(A \cup B)$. Now $N(A \cup B) = N(A) + N(B) - N(A \cap B)$ by the principle of inclusion and exclusion. And $(A \cup B)' = A' \cap B'$. Thus

$$N(A' \cap B') = N - N(A) - N(B) + N(A \cap B) \qquad (ii)$$

Similarly,

$$N(A' \cap B' \cap C') = N - N(A) - N(B) - N(C) + N(A \cap B)$$
$$+ N(A \cap C) + N(B \cap C) - N(A \cap B \cap C) \qquad (iii)$$

We may consider (i), (ii), and (iii) also as the inclusion–exclusion rule involving one subset, two subsets, and three subsets of a set with N elements, respectively.

Now suppose that a_i $(i = 1, 2, 3)$ are three distinct properties associated with the elements of the set X such that a typical element may possess one or more of these properties or may have none of them. Let A_i be the set of all x

in X such that x has property a_i. Let $N(a_i)$ be the number of elements in X with property a_i, $N(a_i')$ be the number of elements in X that do not have the property a_i and $N(a_1, a_2)$ be the number of elements in X possessing both property a_1 and a_2, and so on. Then $N(a_i) = N(A_i)$, $N(a_i, a_j) = N(A_i \cap A_j)$, and $N(a_i, a_j') = N(A_i \cap A_j')$. The inclusion–exclusion rule (iii) given above involving three subsets of X can be rewritten as follows:

$$N(a_1', a_2', a_3') = N - [N(a_1) + N(a_2) + N(a_3)]$$
$$+ [N(a_1, a_2) + N(a_1, a_3) + N(a_2, a_3)] - N(a_1, a_2, a_3)$$

This result is now extended to the case involving n distinct properties (which the elements in a finite set may have) as a theorem that can be proved using a combinatorial argument. Before starting this generalization let us introduce the following notation. Let A_i ($i = 1, 2, \ldots, n$) be n subsets of X. A k-tuple intersection is the intersection of any k distinct subsets of these n sets. The number of k-tuple intersections is, of course, $C(n, k)$. Let S_k be the sum of the number of elements of all the k-tuple intersections. Thus

$$S_1 = N(A_1) + N(A_2) + \cdots + N(A_n)$$
$$S_2 = N(A_1 \cap A_2) + N(A_1 \cap A_3) + \cdots + N(A_{n-1} \cap A_n)$$

and so on.

THEOREM 1.6.1   (The Inclusion–Exclusion Formula)

$$N(A_1' \cap A_2' \cap A_3' \cap \cdots \cap A_n') = N - S_1 + S_2 - S_3 + \cdots + (-1)^n S_n$$

Proof:

For any element x in X and for any subset A of X, the count of x in $N(A)$ is 1 if x is in A. Otherwise, the count is 0. So it is enough if we prove that the count of any element x in X is the same on both sides of the equation.

(a) Suppose that x is not in any one of the n sets. Then the count of x on the left-hand side (LHS) is exactly 1. And this x has a count 1 on the right-hand side (RHS) because x is one of the N elements of X, and it is not in any one of the n sets. Thus the count of any x that is not in one of the n sets is 1 on both sides.

(b) Suppose that x is in exactly one of the n sets. Then the count of this x on the left-hand side is 0. The count of x on the right-hand side is computed as follows: Count of x in $N = 1$, count of X in $S_1 = 1$, and count of $S_i = 0$ when i is not equal to 1. Thus count of x on the right-hand side is $1 - 1 = 0$. More generally, let x be an element that is common to r of the n sets. The count of x on the left-hand side is, of course, 0. Count of x in $N = 1$, count of x in $S_1 = C(r, 1)$, count of x in $S_2 = C(r, 2)$, and so on. Thus the count of x on the right-hand side

$$= 1 - C(r, 1) + C(r, 2) - \cdots + (-1)^r C(r, r) = (1 - 1)^r = 0$$

Thus the count of x is the same on both sides of the equation, and this completes the proof.   ◇

Example 1.6.2

Each student in a freshmen dormitory takes at least one of the four introductory courses in biology (B), English (E), history (H), and mathematics (M). There are 6 students who take all four courses. There are 25 students in each of the four courses, 15 students in any two of the four courses, and 10 students in any three of four courses. How many students are there in the dorm?

Solution. Let N be the number of students. Then $S_1 = C(4, 1) \cdot (25) = 100$, $S_2 = C(4, 2) \cdot (15) = 90$, $S_3 = C(4, 3) \cdot (10) = 40$, and $S_4 = C(4, 4) \cdot (6) = 6$. Since each student takes at least one course, $N(B' \cap E' \cap H' \cap M') = 0$. Thus, by the inclusion–exclusion rule, $0 = N - 100 + 90 - 40 + 6$. So $N = 44$.

If x is any positive integer less than or equal to the positive integer n, the number of multiples of x that do not exceed n is the floor of n/x, which by definition is the smallest nonnegative integer less than or equal to n/x. For example, the number of integers less than 1000 and divisible by 11 is 90, since floor of $1000/11 = 90$. The number of integers less than 15 and divisible by 11 is 1, which is the floor of $15/11$.

Example 1.6.3

Let $X = \{1, 2, 3, \ldots, 600\}$. Find the number of positive integers in X that are not divisible by 3 or 5 or 7.

Solution. Let A, B, and C be the sets of integers in X that are divisible by 3, 5, and 7, respectively. Then $N(A) = $ floor of $600/3 = 200$, $N(B) = $ floor of $600/5 = 120$, and $N(C) = $ floor of $600/7 = 85$. So $S_1 = 200 + 120 + 85 = 405$. Next, $N(A \cap B) = $ number of integers in X divisible by $15 = $ floor of $600/15 = 40$. Similarly, $N(A \cap C) = $ floor of $600/21 = 28$ and $N(B \cap C) = $ floor of $600/35 = 17$. Thus $S_2 = 40 + 28 + 17 = 85$. Finally, $S_3 = N(A \cap B \cap C) = $ floor of $600/105 = 5$. Thus $N(A' \cap B' \cap C') = 600 - 405 + 85 - 5 = 275$. So there are 275 numbers in the set that are not divisible by 3 or 5 or 7.

Two integers m and n are **relatively prime** if the only positive divisor they have in common is 1. The cardinality of the set of positive integers less than n and relatively prime to n is called the **totient function** of n and is denoted by $\phi(n)$. For example, $\phi(8)$ is the cardinality of the set $\{1, 3, 5, 7\}$, so it is equal to 4.

Example 1.6.4

Use the inclusion–exclusion rule to compute $\phi(60)$. The distinct prime divisors of 60 are 2, 3, and 5. Let $N(A)$, $N(B)$, and $N(C)$ be the number of integers less than or equal to 60 and divisible by 2, 3, and 5, respectively. Then $N(A) = (60)/2$, $N(B) = (60)/3$, and $N(C) = (60)/5$. Also, $N(A \cap B) = (60)/(2)(3)$, $N(A \cap C) = (60)/(2)(5)$, $N(B \cap C) = (60)/(3)(5)$, and $N(A \cap B \cap C) = (60)/(2)(3)(5)$. Thus

$$\phi(60) = 60 - \left[\frac{(60)}{2} + \frac{(60)}{3} + \frac{(60)}{5} \right]$$

$$+ \left[\frac{(60)}{(2)(3)} + \frac{(60)}{(2)(5)} + \frac{(60)}{(3)(5)} \right] - \frac{(60)}{(2)(3)(5)}$$

$$= (60) \left(1 - \frac{1}{2} \right) \left(1 - \frac{1}{3} \right) \left(1 - \frac{1}{5} \right)$$

$$= \frac{(60)(2 - 1)(3 - 1)(5 - 1)}{30} = 16$$

This example has a straightforward generalization for any arbitrary integer, as follows.

THEOREM 1.6.2

Let n be any positive integer and let p_i ($i = 1, 2, \ldots, k$) be the distinct prime factors of n. Then $\phi(n) = (n/m) \cdot (p_1 - 1) \cdot (p_2 - 1) \cdot \ldots \cdot (p_k - 1)$, where m is the product of the k distinct prime factors of n. $\diamond$

Example 1.6.5

There is a microcomputer in each of the 32 faculty offices in a department. Fifteen of them have color (C) monitors, 10 of them have laser (L) printers, and 8 of them have modems (M). Two of them have all three options. At least how many have none of the three options?

Solution.   If r is the number of offices with none of these options, $r = 32 - [15 + 10 + 8] + [N(C \cap L) + N(C \cap M) + N(L \cap M)] - 2,$

where each unknown number is at least 2. Thus $r \geqslant 32 - 33 + 6 - 2 = 3$.

Example 1.6.6

Find the number r of solutions in integers of the equation $a + b + c = 25$, where a is at least 2 and at most 4, b is at least 3 and at most 6, and c is at least 4 and at most 8.

Solution. The required number r is the number of solutions in integers of the revised equation $x + y + z = 16$, where the upper bounds for x, y, and z are 2, 3, and 4, respectively. Let

$$N = \text{number of solutions of the revised equation}$$

$$X = \text{set of solutions such that } x \text{ is at least 3}$$

$$Y = \text{set of solutions such that } y \text{ is at least 4}$$

$$Z = \text{set of solutions such that } z \text{ is at least 5}$$

Then $N = C(16 + 2, 2)$, $N(X) = C(16 - 3 + 2, 2)$, $N(Y) = C(16 - 4 + 2, 2)$, $N(Z) = C(16 - 5 + 2, 2)$, $N(X \cap Y) = C(16 - 3 - 4 + 2, 2)$, $N(X \cap Z) = C(16 - 3 - 5 + 2, 2)$, $N(Y \cap Z) = C(16 - 4 - 5 + 2, 2)$, and $N(X \cap Y \cap Z) = C(16 - 3 - 4 - 5 + 2, 2)$. Thus

$$r = C(18, 2) - [C(15, 2) + C(14, 2) + C(13, 2)]$$
$$+ [C(11, 2) + C(10, 2) + C(9, 2)] - C(6, 2)$$
$$= 153 - (105 + 91 + 78) + (55 + 45 + 36) - 15 = 0$$

Derangements

Suppose that X is a finite set with n elements and each element in the set is assigned a unique positive integer (a label) between 1 and n. The element that is assigned the label i is the ith element of the set. If in a permutation of these elements, the ith element appears in the ith position, that element is in its **original position** for X. A **derangement** of X is a permutation in which no element appears in its original position. For example, let $X = \{a, b, c, d\}$ and the labels of a, b, c, d be 1, 2, 3, 4, respectively. Then the permutation $abdc$ is not a derangement, because a and b are in their original positions. But the permutation $badc$ is a derangement.

As a practical example, consider the following scenario: A student is sending out applications for employment to various hiring agencies. She completed 10 different applications addressed to 10 different agencies and wrote the addresses of these agencies on 10 identical envelopes. She then told her brother to put each application in the right envelope and mail all the applications to the respective agencies. She has a derangement at hand if no application was inside the right envelope!

The total number of derangements of a set of cardinality n is denoted by D_n.

THEOREM 1.6.3

The total number of derangements of a set of cardinality n is

$$D_n = (n!) \left[1 - \frac{1}{1!} + \frac{1}{2!} - \frac{1}{3!} + \cdots + \frac{(-1)^n}{n!} \right]$$

Proof:

Let N be the total number of permutations on X and let A_i be the set of permutations in which the ith object is in its original place. The total number of permutations on X is $n!$. Thus $D_n = n! - S_1 + S_2 + \cdots + (-1)^n S_n$, where

$$S_1 = C(n, 1) \cdot (n - 1)!$$
$$S_2 = C(n, 2) \cdot (n - 2)!$$
$$\vdots$$
$$S_k = C(n, k) \cdot (n - k)!$$

and so on. So

$$D_n = (n!) \left[1 - \frac{1}{1!} + \frac{1}{2!} - \frac{1}{3!} + \cdots + \frac{(-1)^n}{n!} \right]$$

Using the principle of inclusion and exclusion, we can obtain some important results involving the number of r-sequences and the number of partitions of a finite set. The following two theorems are direct consequences of Theorem 1.6.1. We need a notation before these theorems are presented.

If n and r are two positive integers $(n \leq r)$, the **Stirling number of the second kind,** denoted by $S(r, n)$, is defined by the following relation:

$$(n!) \cdot S(r, n) = n^r - C(n, 1) \cdot (n - 1)^r + C(n, 2) \cdot (n - 2)^r - \cdots$$
$$+ (-1)^{n-1} C(n, n - 1) \cdot 1^r$$

THEOREM 1.6.4

The number of r-sequences that can be formed using the elements of a set with n elements such that in every such sequence each element of the set appears at least once is $(n!) \cdot S(r, n)$.

Proof:

Let $X = \{x_i : i = 1, 2, \ldots, n\}$ and A_i be the set of r-sequences that do not contain x_i. Then $S_i = C(n, i) \cdot (n - i)^r$. The stated result is an immediate consequence of Theorem 1.6.1.   ◇

THEOREM 1.6.5

(a) *The Allocation Problem.* The number of ways of allocating r distinct objects to n locations such that each location receives at least one object is $(n!) \cdot S(r, n)$.

(b) *The Set Partitioning Problem.* (1) The number of partitions of a set of cardinality r such that each partition has n nonempty sets is $S(r, n)$ and (2) the number of partitions of a set with r elements such that each partition has at most n nonempty sets is $S(r, n) + S(r, n - 1) + \cdots + S(r, 1)$.

Proof:

These two results follow from Theorem 1.6.1 and the definition of $S(r, n)$.   ◇

[Notice the difference between the assertions in Theorem 1.3.5 and Theorem 1.6.5(b).]

Number of Functions from a Finite Set to Another Finite Set

Let X and Y be two finite sets with cardinality r and n, respectively. Suppose that f is any arbitrary function from X to Y. Any one of the r elements from X can be mapped into any one of the n elements of Y in n ways. So by the multiplication rule there are n^r functions from X to Y. But if f is an *injection*, the number of ways the r elements can be mapped is much less: It is, in fact, equal to $n(n - 1)(n - 2) \cdots (n - r + 1)$. If the mapping is a *surjection*, every element in Y has a preimage. So by applying Theorem 1.6.5 (a), we see that the number of surjections is $(n!) \cdot S(r, n)$. We can summarize these results as a theorem.

THEOREM 1.6.6

Let X be a set of cardinality r and Y be a set of cardinality n. Then there are (1) n^r functions from X to Y, (2) $P(n, r)$ injections from X to Y, and (3) $(n!) \cdot S(r, n)$ surjections from X to Y.

A Generalization of the Inclusion–Exclusion Principle

(This part may be omitted without loss of continuity.)

We conclude this section with a theorem that is a generalization of Theorem 1.6.1. For this we need some notations. With each element x of a set X with N elements, we associate a nonnegative real number $w(x)$ called the *weight of the element*. We have a set P of n properties. A typical element in X may or may not have some or all the properties listed in P. $V(r)$ is the sum of the weights of all elements x where x satisfies *exactly* r of these n properties. $U(r)$ is the sum of the weights of all elements x where x satisfies *at least* r of these n properties. For example, let a, b, c, d, e the elements of a set X with weights 6, 7, 8, 9, 10, respectively. P is a set of four properties: s, t, u, v. It is known that a satisfies s and t; b satisfies s, u, and v; c satisfies t and u; d satisfies s, t, and u; and e satisfies all the four properties. Then $V(1) = 0$, $V(2) = 6 + 8 = 14$, $V(3) = 7 + 9 = 16$, and $V(4) = 10$; and $U(1) = 6 + 7 + 8 + 9 + 10 = 40$, $U(2) = 40$, $U(3) = 26$ and $U(4) = 10$. Finally, if Q is a subset of P, we define the weight $W(Q)$ of the set Q as the sum of the weights of all elements x where x satisfies all the r properties listed in Q, and $W(r)$ is the sum of all terms of the type $W(Q)$, where Q is a subset of P with r elements. For example, if $Q = \{s, t\}$, then $W(Q) = w(a) + w(d) + w(e) = 25$. $W(2)$ is the sum of the weights of all the two-element subsets of P. In this example P has six subsets of cardinality 2. Thus $W(2) = 25 + 26 + 17 + 27 + 10 + 17 = 122$. $W(0)$ by definition is the sum of the weights of all the N elements in X.

THEOREM 1.6.7   (Generalized Inclusion–Exclusion Formula)

(a) $V(r) = W(r) - C(r + 1, r)W(r + 1) + C(r + 2, r)W(r + 2) - \cdots + (-1)^{n-r}C(n, r)W(n)$. (Here $r \geq 1$.)   (*)

(b) $U(r) = W(r) - C(r, r - 1)W(r + 1) + C(r + 1, r - 1)W(r + 2) - \cdots + (-1)^{n-r}C(n - 1, r - 1)W(n)$. (Here $r > 1$.)

Proof:

(a) Obviously, a typical element x contributes its weight $w(x)$ to the left-hand side of (*) if and only if x satisfies exactly r of the n properties. So it is enough if we prove that x contributes $w(x)$ to the right-hand side if and only if x satisfies exactly r of the n properties.

Suppose that x satisfies s of the n properties. If $s = r$, the element x contributes $w(x)$ to $W(r)$ in the right-hand side and 0 to the other terms. If $s < r$, the contribution of x to the right-hand side is 0. It remains to be proved that the contribution of x to the right-hand side is 0 when $s > r$. In this case, the contribution of x to $W(r)$ is $C(s, r)w(x)$. The contribution of x to $W(r + 1)$ is $C(s, r + 1)w(x)$, and so on. Thus when $s > r$, the contribution of x to the right-hand side is

$$w(x)[C(s, r) - C(s, r + 1)C(r + 1, r) + C(s, r + 2)C(r + 2, r)$$
$$- \cdots + (-1)^{s-r}C(s, s)C(s, r)]$$

Now it is easily verified that

$$C(i, j)C(j, k) = C(i, k)C(i - k, i - j)$$

Thus the contribution of x to right-hand side is $w(x) \cdot C(s, r) \cdot K$ where

$$K = 1 - C(s - r, s - r - 1) + C(s - r, s - r - 2) - \cdots$$
$$= 1 - C(s - r, 1) + C(s - r, 2) - \cdots$$
$$= (1 - 1)^{s-r} = 0$$

This completes the proof of (a). If we put $r = 0$ and $w(x) = 1$ for each element x in X, we get the formula in Theorem 1.6.1 as a special case.

(b) This is left as an exercise.   ◇

Example 1.6.7

Verify the formula given in Theorem 1.6.7 for the following data: The elements of a set X are a, b, c, d, and e with weights 6, 7, 8, 9, and 10. P is a set of four properties s, t, u, v. It is known that a satisfies p and q; b satisfies s, u, and v; c satisfies t and u; d satisfies s, t, and u; and finally, e satisfies all the four properties.

Solution. We have already computed the following: $V(1) = 0$, $V(2) = 14$, $V(3) = 16$, $V(4) = 10$, $U(1) = 40$, $U(2) = 40$, $U(3) = 26$, $U(4) = 10$, and $W(2) = 122$. Next we find that $W(1) = 32 + 33 + 34 + 17 = 116$, $W(3) = 19 + 10 + 17 + 10 = 56$, and $W(4) = 10$.

Formula (i) when $r = 1$:

$$\begin{cases} \text{LHS} = V(1) = 0 \\ \text{RHS} = W(1) - C(2, 1)W(2) + C(3, 1)W(3) - C(4, 1)W(4) \\ \qquad\quad = 116 - 244 + 168 - 40 = 0 \end{cases}$$

Formula (i) when $r = 2$:

$$\begin{cases} \text{LHS} = V(2) = 14 \\ \text{RHS} = 122 - (3)(56) + (6)(10) = 14 \end{cases}$$

Formula (i) when $r = 3$:

$$\begin{cases} \text{LHS} = V(3) = 16 \\ \text{RHS} = 56 - (4)(10) = 16 \end{cases}$$

Formula (i) when $r = 4$:

$$\begin{cases} \text{LHS} = V(4) = 10 \\ \text{RHS} = 10 \end{cases}$$

Formula (ii) when $r = 2$:

$$\begin{cases} \text{LHS} = U(2) = 40 \\ \text{RHS} = 122 - (2)(56) + (3)(10) = 40 \end{cases}$$

Formula (ii) when $r = 3$:

$$\begin{cases} \text{LHS} = U(3) = 26 \\ \text{RHS} = 56 - (3)(10) = 26 \end{cases}$$

Formula (ii) when $r = 4$:

$$\begin{cases} \text{LHS} = U(4) = 10 \\ \text{RHS} = 10 \end{cases}$$

1.7 SUMMARY OF RESULTS IN COMBINATORICS

We now conclude this chapter with a complete list of all the important results involving permutations, combinations, allocations, derangements, set partitions, and number of mappings between finite sets established in these pages.

1. A permutation is a linear arrangement of objects where the order in which distinct objects appear is crucial, whereas a combination is just a collection

of objects and the order in which objects are chosen for inclusion in it is not relevant.

2. (a) The number of r-permutations with r distinct elements that can be formed using the elements of a collection of n distinct elements, (b) the number of ways of allocating r *distinct* objects to n locations such that no location can receive more than one object, and (c) the number of injections from a set of r elements to a set of n elements are all equal to $P(n, r) = n!/(n - r)!$

3. (a) The number of r-combinations with r distinct elements that can be formed using the elements of a collection of n distinct elements and (b) the number of ways of allocating r *identical* objects to n locations such that no location can receive more than one object are both equal to $C(n, r) = n!/[r!(n - r)!] = C(n, n - r)$.

4. The coefficient of x^r in $(1 + x)^n$ is $C(n, r)$.

5. (a) The number of r-sequences with r elements (not necessarily distinct) that can be formed using the elements of a collection of n distinct elements, (b) the number of ways of placing r *distinct* objects in n locations with no restriction on the number of objects a location can receive, and (c) the number of functions from a set of r elements to a set of n elements are all equal to n^r.

6. (a) The number of r-collections with r elements (not necessarily distinct) that can be formed using the elements of a collection of n distinct elements, (b) the number of ways of allocating r *identical* objects to n locations with no restriction on the number of objects a location can receive, (c) the number of solutions in nonnegative integers of $x_1 + x_2 + \cdots + x_n = r$, and (d) the number of terms in the expansion of $(x_1 + x_2 + \cdots + x_n)^r$ are all equal to $C(r + n - 1, n - 1)$.

7. If $p_1, p_2, \ldots, p_n$ are nonnegative integers whose sum is p, then (a) the number of r-collections with r elements (not necessarily distinct) that can be formed using the n distinct elements of $X = \{x_1, x_2, \ldots, x_n\}$ such that in each combination x_1 appears at least p_i times ($i = 1, 2, \ldots, n$), (b) the number of ways of allocating r *identical* objects to n locations such that location i gets at least p_i objects ($i = 1, 2, \ldots, n$), and (c) the number of solutions in nonnegative integers of $x_1 + x_2 + \cdots + x_n = r$ where x_i is at least p_1 ($i = 1, 2, \ldots, n$) are all equal to $C(r - p + n - 1, n - 1)$.

8. Define $P(t; t_1, t_2, \ldots, t_j) = P(t, s)/[(t_1!)(t_2!) \cdots (t_j!)]$ where $s = t_1 + t_2 + \cdots + t_j$.

9. If there are n objects of k different types such that the objects in each type are identical and if type i has n_i objects ($i = 1, 2, \ldots, k$) then there are $P(n; n_1, n_2, \ldots, n_k)$ ways of arranging these n objects in a line.

10. If there are k types of objects and if type i has n_i identical objects ($i = 1$, $2, \ldots, k$), then there are $P(n; n_1, n_2, \ldots, n_k)$ ways of allocating these $n_1 + n_2 + \cdots + n_k$ objects to n locations such that no location can receive more than one object.

11. There are $P(n; n_1, n_2, \ldots, n_k)$ ways of allocating n *distinct* objects to k locations such that location i gets exactly n_i objects for each $i = 1, 2,$ $\ldots, k$.

12. The coefficient of $x_1^{n_1} x_2^{n_2} \cdots x_k^{n_k}$ in $(x_1 + x_2 + \cdots + x_k)^n$ is $P(n; n_1, n_2, \ldots, n_k)$.

13. A set of cardinality n can be partitioned into a class consisting of p_1 subsets each of cardinality n_1, p_2 subsets each of cardinality $n_2, \ldots,$ and p_k subsets each of cardinality n_k in $(n!)/\{[(p_1!)(n_1!)^{p_1}][(p_2!)(n_2!)^{p_2}] \cdot \cdots \cdot [(p_k!)(n_k!)^{p_k}]\}$ ways where the integers $n_1, n_2, \ldots, n_k$ are distinct.

14. When n and r are positive integers, the Stirling number of the second kind, denoted by $S(r, n)$, is defined as $(n!) \cdot S(r, n) = n^r - C(n, 1) \cdot (n - 1)^r + C(n, 2) \cdot (n - 2)^r + \cdots + (-1)^{n-1}C(n, n - 1) \cdot 1^r$.

15. The number of r-sequences (elements not necessarily distinct) that can be formed using the elements of a collection X of n distinct elements such that in every permutation each element of X appears at least once is $(n!) \cdot S(r, n)$.

16. The number of ways of allocating r *distinct* objects to n locations such that every location receives at least one object is $(n!) \cdot S(r, n)$.

17. The number of partitions of a set with r elements such that each partition has n nonempty sets is $S(r, n)$.

18. The number of partitions of a set with r elements such that each partition has at most n nonempty sets is $S(r, n) + S(r, n - 1) + \cdots + S(r, 1)$.

19. The number of surjections from a set of r elements to a set of n elements is $(n!) \cdot S(r, n)$.

20. The number of derangements of a set with n elements is
$D_n = (n!)[1 - 1/1! + 1/2! - 1/3! + \cdots + (-1)^n/n!]$.

1.8 NOTES AND REFERENCES

Combinatorics is one of the most venerable branches of mathematics. Formulas involving arrangements, sequences, and combinations were known to the Chinese, Hindu, and Greek mathematicians as early as the first century A.D. Combinatorics is tied very closely with probability theory, and in the seventeenth and eighteenth centuries many European mathematicians were interested in the study of combinatorial probability.

Some excellent general references in the area of combinatorics are the books by Aigner (1979), Anderson (1979), Cohen (1978), Krishnamurthy (1986), Liu (1968), Riordan (1978), Roberts (1984), and Tucker (1984). There is also the classic text by MacMahon (1960). The first comprehensive book dealing with permutations and combinations is by Whitworth (1901). Chapter 1 of Grimaldi (1985), Chapter 3 of Liu (1985), and Chapter 2 of Townsend (1987) also deal with the material of this chapter. Algorithms for generating permutations and combinations of a given finite set are given in detail in Chapters 1 and 2 of Even (1973) and in Chapter 5 of Reingold et al. (1977).

One does not have to be a mathematician to know that if there are more objects (pigeons) than containers (pigeonholes), there will be at least one container with two or more objects. Notice that this is an existential statement: It simply asserts that there *exists* a container with at least two objects. Neither the container nor the objects are identifiable. It goes on record that it was Gustav Dirichlet (1805–1858) who used this principle extensively in his investigation of problems in number theory—hence the name Dirichlet pigeonhole principle, which is also known as the 'shoebox principle.' The nontrivial generalizations of this deceptively innocuous principle involve some of the most profound and deep results in all of combinatorial theory. Example 1.5.5 is a very special case of a result known as Ramsey's theorem. References to Ramsey theory include Chapter 5 of Cohen (1978), Chapter 8 of Roberts (1984), Chapter 4 of Ryser (1963), and the book on Ramsey theory by Graham et al. (1980).

The pioneering work using the inclusion–exclusion principle was done by James Sylvester (1814–1897) and its importance and usefulness were made public with the publication of the book *Choice and Chance* by Whitworth (1901). For a discussion of this principle, refer to Chapter 5 of Anderson (1979), Chapter 5 of Cohen (1978), Chapter 7 of Grimaldi (1985), Chapter 4 of Liu (1968), Chapter 3 of Riordan (1978), Chapter 6 of Roberts (1984), Chapter 2 of Ryser (1963), and Chapter 8 of Tucker (1984).

1.9 EXERCISES

1.1. The social security number of a person is a sequence of nine digits that are not necessarily distinct. If X is the set of all social security numbers, find the number of elements in X.

1.2. There are six characters—three letters of the English alphabet followed by three digits—which appear on the back panel of a particular brand of a printer as an identification number. If X is the set of all possible identification numbers for this brand of printer, find the number of elements in X if (a) characters can repeat in an identification number, (b) digits cannot repeat, (c) letters cannot repeat, and (d) characters cannot repeat.

1.3. Find the number of ways of picking (a) a king and a queen, (b) a king or a queen, (c) a king and a red card, and (d) a king or a red card from a deck of cards.

1.4. **(a)** Find the number of even numbers between 0 and 100. **(b)** Find the number of even numbers with distinct digits between 0 and 100.

1.5. A sequence of digits where each digit is 0 or 1 is called a *binary number*. Each digit in a binary number is a component of the number. A binary number with eight components is called a **byte**. **(a)** Find the number of bytes. **(b)** Find the number of bytes that begin with 10 and end with 01. **(c)** Find the number of bytes that begin with 10 but do not end with 01. **(d)** Find the number of bytes that begin with 10 or end with 01.

1.6. A variable name in the programming language BASIC is either a letter of the alphabet or a letter followed by a digit. Find the number of distinct variable names in this language.

1.7. A sequence of characters is called a **palindrome** if it reads the same way forward or backward. For example, 59AA95 is a six-character palindrome, and 59A95 is a five-character palindrome. Some other instances of palindromes: U NU, LON NOL, MALAYALAM, NOW ON, PUT UP, TOO HOT TO HOOT, NEVER ODD OR EVEN, ABLE WAS I ERE I SAW ELBA, and POOR DAN IS IN A DROOP. Find the number of nine-character palindromes that can be formed using the letters of the alphabet such that no letter appears more than twice in each of them.

1.8. Find the number of ways to form a four-letter sequence using the letters A, B, C, D, and E if **(a)** repetitions of letters are permitted, **(b)** repetitions are not permitted, **(c)** the sequence contains the letter A but repetitions are not permitted, and **(d)** the sequence contains the letter A but repetitions are permitted.

1.9. There are n married couples in a group. Find the number of ways of selecting a woman and a man who is not her husband from this group.

1.10. Let X be the set of all polynomials of degree 4 in a single variable t such that every coefficient is a single-digit nonnegative integer. Find the cardinality of X.

1.11. A variable name in the programming language FORTRAN is a sequence that has at most six characters such that the first character is a letter of the alphabet and the remaining characters, if any, are either letters or digits. Find the number of distinct variable names in this language.

1.12. There are 10 members—A, B, C, D, E, F, G, H, I, and J—in a fund-raising committee. The first task of the committee is to choose a chairperson, a secretary, and a treasurer from this group. No individual can hold more than one office. Find the number of ways of selecting a chairperson, a secretary, and a treasurer such that **(a)** no one has any objection for holding any of these three offices, **(b)** C would like to be the chairperson, **(c)** B would not like to be the chairperson, **(d)** A does not like to be either the chairperson or the secretary, **(e)** I or J would like to be the treasurer, and **(f)** E or F or G would like to hold one of these three offices.

1.13. There are three bridges connecting two towns, A and B. Between towns B and C there are four bridges. A salesperson has to travel from A to C via B. Find **(a)** the number of possible choices of bridges from A to C, **(b)** the number of choices for a round-trip travel from A to C, and **(c)** the number of choices for a round-trip travel if no bridge is repeated.

1.14. Compute **(a)** $P(8, 5)$, **(b)** $P(9, 2)$, and **(c)** $P(6, 6)$.

1.15. Prove Theorem 1.2.2.

1.16. Find the value of the positive integer n if **(a)** $P(n, 2) = 30$, **(b)** $P(n, 3) = 24 \cdot P(n, 2)$, and **(c)** $10 \cdot P(n, 2) = P(3n - 1, 2) + 40$.

1.17. Compute 6!. Use this result to compute 7! and 8!.

1.18. A and B are two members in a party of 12. Find the number of ways of assigning these 12 people to 12 rooms situated in a row such that each person gets a room and **(a)** A and B are next to each other and **(b)** A and B are not next to each other.

1.19. Show that $P(n, r + 1) = (n - r) \cdot P(n, r)$ and use this result to find the value of n if $P(n, 9) = 15 \cdot P(n, 8)$.

1.20. Find the value of k if $P(n + 1, r) = k \cdot P(n, r)$. Use this result to find n and r if $k = 5$, $n > r$, and r is as small as possible.

1.21. Four station wagons, five sedans, and six vans are to be parked in a row of 15 parking spots. Find the number of ways of parking these vehicles such that **(a)** the station wagons are parked at the beginning, then the sedans, and then the vans, and **(b)** vehicles of the same type are parked en bloc.

1.22. Consider a collection of six stones of different colors: blue (B), green (G), pink (P), red (R), white (W), and yellow (Y). Find **(a)** the number of ways of making a tiepin on which these stones are to be placed in a row, **(b)** the number of ways of making a brooch on which these six stones are to be mounted in a circular pattern, and **(c)** the number of ways of making a ring using these six stones.

1.23. Eight people are to be seated around a large round table for a conference. Find the number of possible seating arrangements.

1.24. A mother and her two small children join seven members of her family for dinner and they have to sit around a round table. Find the number of possible seating arrangements so that the two children can sit on either side of the mother.

1.25. Six girls and six boys are to be assigned to stand around a circular fountain. Find the number of such assignments if on either side of a boy there is a girl and on either side of a girl there is a boy.

1.26. If X and Y are two sets with n elements each and if there are no elements common to the two sets, find the number of ways arranging the $2n$ elements of these two sets in a circular pattern so that on either side of an element of X there is an element of Y, and vice versa.

1.27. Compute **(a)** $P(10; 4, 4, 2)$ and **(b)** $P(12; 5, 4, 3)$.

1.28. Compute **(a)** $P(17; 4, 3, 2)$ and **(b)** $P(17; 2, 2, 2)$.

1.29. Prove that if m and n are positive integers, $(mn)!/(m!)^n$ is also a positive integer.

1.30. Find the number of ways in which the complete collection of letters that form the word MISSISSIPPI can be arranged such that **(a)** there is no restriction on the location of the letters, and **(b)** all the S's stay together.

1.31. Find the number of ways of **(a)** assigning 9 students to 11 rooms (numbered serially from 100 to 110) in a dormitory so that each room has at most one occupant, and

(b) installing nine color telephones (two red, three white, and four blue) in these rooms, so that each room has at most one telephone.

1.32. Compute (a) $C(9, 4)$, (b) $C(10, 7)$, and (c) $C(8, 4)$

1.33. X is a set with nine elements. Find the number of (a) subsets of X, (b) subsets of cardinality 3, and (c) unordered pairs in X.

1.34. Prove Pascal's formula algebraically.

1.35. There are 4 women and 9 men in the mathematics faculty of a college. Find the number of ways of forming a hiring committee consisting of 2 women and 3 men from the department.

1.36. There are 5 distinct white and 7 distinct blue shirts in a wardrobe. Find the number of ways of taking 4 shirts from the wardrobe such that (a) they could be either white or blue, (b) they are all white, (c) they are all blue, and (d) they are all of the same color, and (e) 2 are white and 2 are blue.

1.37. Find the number of ways of seating r people from a group of n people around a round table.

1.38. Find the number of ways of seating 14 people such that 8 of them are around one round table and the rest are around another round table.

1.39. Find the number of ways of seating 14 people such that 8 of them are around a round table and the rest are on a bench.

1.40. Find the number of bytes that can be formed using exactly six zeros.

1.41. Find the number of ways in which the letters that appear in MISSISSIPPI can be rearranged so that no two S's are adjacent.

1.42. In a state lottery, a ticket consists of six distinct integers chosen from the set $X = \{1, 2, 3, \ldots, 42\}$. On every Saturday at 8:00 P.M., six distinct integers are chosen from X by a computer. A ticket buyer wins (1) the first prize (the jackpot) if the six numbers in the ticket are the same as the six numbers picked by the computer, (2) the second prize if any five numbers in the ticket are picked by the computer, (3) the third prize if any of the four numbers in the ticket are picked by the computer, and (4) the fourth prize if any of the three numbers in the ticket are picked by the computer. Find (a) the number of distinct tickets that one has to buy which will definitely assure the buyer winning the jackpot and the probability of winning the jackpot if a person buys 1000 tickets, (b) the probability of winning the second prize if a person buys a single ticket, and (c) the probability of winning the third prize if the person buys a single ticket.

1.43. Prove the following identity using a combinatorial argument:

$$C(n, r) = C(r, r) \cdot C(n - r, 0) + C(r, r - 1) \cdot C(n - r, 1)$$
$$+ C(r, r - 2) \cdot C(n - r, 2) + C(r, r - 3) \cdot C(n - r, 3) + \cdots$$
$$+ C(r, 1) \cdot C(n - r, r - 1) + C(r, 0) \cdot C(n - r, r)$$

1.44. If $C(n, r) = C(r, 1) \cdot C(n, r - 1)$, solve n in terms of r.

1.45. Prove that $C(pn, pn - n)$ is a multiple of p.

1.46. Prove the identity $C(3n, 3) = 3C(n, 3) + 6n \cdot C(n, 2) + n^3$ using a combinatorial argument.

1.47. Let X be the set of all words of length 10 in which the letter P appears 2 times, Q appears 3 times, and R appears 4 times. Find the cardinality of X.

1.48. A mother bought 10 story books for her 3 children. The youngest gets 2 books and the other two get 4 each. Find the number of ways she can pack them as gifts.

1.49. A linear algebra class consists of 10 mathematics majors and 12 computer science majors. A team of 12 has to be selected from this class. Find the number of ways of selecting a team if **(a)** the team has 6 from each discipline, and **(b)** the team has a majority of computer science majors.

1.50. Find the coefficient of $a^2b^3c^3d^4$ in the expansion of **(a)** $(a + b + c + d)^{12}$ and **(b)** $(2a - 3b + 2c - d)^{12}$.

1.51. Use Pascal's triangle and list the coefficients of the terms which appear in the expansion of $(x + y)^n$ when $n = 4, 5$, and 6.

1.52. Use a combinatorial argument to prove **Newton's identity:** $C(n, r) \cdot C(r, k) = C(n, k) \cdot C(n - k, r - k)$.

1.53. Prove the following identity:

$$C(n, 0) + C(n + 1, 1) + C(n + 2, 2)$$

$$+ \cdots + C(n + r, r) = C(n + r + 1, r)$$

1.54. Prove: $C(n, 0) + C(n, 1) + C(n, 2) + \cdots + C(n, n) = 2^n$.

1.55. Use a combinatorial argument to prove the following:

$$[C(n, 0)]^2 + [C(n, 1)]^2 + [C(n, 2)]^2 + \cdots + [C(n, n)]^2 = C(2n, n)$$

1.56. Prove the following identity:

$$C(m, 0), C(n, 0) + C(m, 1), C(n, 1) + C(m, 2) \cdot C(n, 2)$$

$$+ \cdots + C(m, n) \cdot C(n, n) = C(m + n, n)$$

1.57. There are 18 students in a class. Find the number of ways of partitioning the class into **(a)** 4 groups of equal strength and a minority group, **(b)** 2 groups of 5 students, 1 group of 4 students, and 2 groups of 2 students, and **(c)** 1 group of 7 students, 1 group of 6 students, and 1 group of 5 students.

1.58. Find the number of r-sequences that can be formed using the elements of the set $X = \{A, B, C, D, E, F, G\}$ if **(a)** $r = 4$ and the elements in each sequence is distinct, **(b)** $r = 4$, and **(c)** $r = 9$.

1.59. Find the number of r-collections that can be formed using the elements of the set $X = \{A, B, C, D, E, F, G\}$ if **(a)** $r = 4$ and the elements in each collection are distinct, **(b)** $r = 4$, and **(c)** $r = 9$.

1.60. Find the number of distinct solutions in nonnegative integers of the equation $a + b + c + d + e = 24$.

1.61. Find the number of terms in the multinomial expansion of $(a + b + c + d + e)^{24}$.

1.62. Find the number of ways of forming a team of 15 students from a large university to represent freshmen, sophomores, juniors, seniors, and graduate students such that the team has **(a)** at least one from each group, **(b)** at least two from each group, and **(c)** at least two graduate students.

1.63. Find the number of solutions of the linear equation $a + b + c + d + e = 10$ if **(a)** all the variables are nonnegative integers, **(b)** all the variables are positive integers, and **(c)** all the variables are positive integers and the variable a is odd.

1.64. Find the number of ways a mother can distribute 9 identical candy bars to her three children so that each child gets at least 2 bars.

1.65. The sum of the four positive integers a, b, c, and d is at most 10. Find the number of possible choices for these integers.

1.66. When a die is rolled, one of the first six positive integers is obtained. Suppose that the die is rolled five times and the sum of the five integers thus obtained is added. The five throws constitute a trial. Find the number of possible trials such that the sum is at most 12.

1.67. Establish the following identity:

$$C(n, n) + C(n + 1, n) + C(n + 2, n) + \cdots + C(n + r, n)$$

$$= C(n + r + 1, n + 1)$$

1.68. Find the number of solutions in nonnegative integers of the equation $x_1 + x_2 + x_3 + 3x_4 = 7$.

1.69. If $X = \{x_1, x_2, \ldots, x_n\}$ is a collection of n distinct objects and r any positive integer, find the number of r-collections of X such that each such collection has the object x_i repeated at least p_i times where $i = 1, 2, 3, \ldots, n$.

1.70. Find the number of ways of allocating r identical objects to n distinct locations such that location i gets at least p_i objects, where $i = 1, 2, \ldots, n$.

1.71. Find the number of solutions in nonnegative integers of the (strict) inequality $a + b + c + d + e < 11$.

1.72. Solve Problem 1.71 if a is at most 6.

1.73. Show that it is possible to have a set of 5 people such that there is no subgroup of 3 strangers or a subgroup of 3 people known to one another in this set.

1.74. There are 4 commuter flights from city A to city B daily. For a particular day, it was noticed that the number of vacant seats on these flights are 8, 10, 13, and 9, respectively. Find the minimum number of tickets that have to be sold so that the number of vacant seats will be **(a)** at most 1 in flight 1 or at most 3 in flight 2 or at most 6 in flight 3 or at most 2 in flight 4, **(b)** at most 2 in flight 1 or at most 3 in flight 2 or at most 4 in flight 3 or at most 1 in flight 4.

1.75. Prove that in any group of 10 people either there is a subgroup of 3 strangers or a subgroup of 4 people known to one another.

1.76. The numbers $1, 2, 3, \ldots, n$ (n is at least 3) are randomly placed around a circle and r is any integer less than n. Let S_i be the sum of the r consecutive integers (considered clockwise) starting from i and including i, where $i = 1, 2, \ldots, n$. Show that there is at least one S_i that is not smaller than the floor of $r(n + 1)/2$.

1.77. Show that in every finite set of numbers there is a number that is greater than or equal to the arithmetic mean of the numbers in the set.

1.78. Let $X = \{1, 2, 3, \ldots, 600\}$. Find the number of elements in X that are not divisible by 3 or 5 or 7.

1.79. (a) Obtain a formula to find the number of primes not exceeding a given positive integer. (b) Use this formula to find the number of primes not exceeding 100.

1.80. A positive integer is **squarefree** if it is not divisible by the square of an integer greater than 1. (a) Obtain a formula to compute the number of squarefree integers not exceeding a given positive integer, and (b) use this formula to compute the number of squarefree numbers not exceeding 100.

1.81. If p and q are two distinct primes, find the totient function of pq.

1.82. There are six chairs marked 1 to 6 in the conference room of an office. Six people attend a seminar in this room in the morning and again in the afternoon. (a) Find the number of permutations and derangements regarding the seating arrangements, (b) find the probability that nobody sits in the same seat twice, (c) find the probability that exactly one person sits in the same chair twice, (d) find the probability that at least one person gets the same seat twice, (e) find the probability that exactly two people retain their seats, and (f) find the probability that all the six retain their seats.

1.83. Use a combinatorial argument to establish the identity:

$$C(n, 0) \cdot D_n + C(n, 1) \cdot D_{n-1} + C(n, 2) \cdot D_{n-2} + \cdots + C(n, n) \cdot D_0 = n!.$$

1.84. Find the number of solutions in integers of the linear equation $p + q + r = 25$ where p is at least 2 and at most 4, q is at least 3 and at most 6, and r is at least 4 and at most 8.

1.85. Let X be the set of 4-sequences that can be formed using the letters A, B, C, D, E, and F such that every sequence in X has the letters A, B, and C at least once. Find the cardinality of X.

1.86. There are 5 job openings in an office. On the basis of a written test and a personal interview, 4 candidates were selected and each candidate is offered one of the available jobs. Find the number of ways of assigning these jobs to the candidates.

1.87. Find the number of permutations of the nine digits $1, 2, \ldots, 9$ in which (a) the blocks 12, 34, and 567 do not appear, and (b) the blocks 12, 23, and 415 do not appear.

1.88. Let $D(n, r)$ be the number of permutations of a set of n elements in which exactly r of the n elements appear in their "natural" positions and $E(n, r)$ be the number of permutations in which at least r of the n elements appear in their natural positions. Prove **(a)** $D(n, 0) = D_n$, **(b)** $D(n, r) = C(n, r) \cdot D_{n-r}$, **(c)** $D(n, n) = E(n, n) = 1$, and **(d)** if $S(i) = C(n, i) \cdot (n - i)!$, then

$$D(n, r) = S(r) - C(r + 1, r) \cdot S(r + 1) + C(r + 2, r) \cdot S(r + 2)$$
$$- \cdots + (-1)^{n-r}C(n, r) \cdot S(n)$$

$$E(n, r) = S(r) - C(r, r - 1) \cdot S(r + 1) + C(r + 1, r - 1) \cdot S(r + 2)$$
$$- \cdots + (-1)^{n-r}C(n - 1, r - 1) \cdot S(n)$$

CHAPTER

2

Generating Functions

2.1 INTRODUCTION

In this chapter we introduce the concept of a generating function—a powerful tool that is very useful in solving counting problems, particularly problems involving the selection and arrangement of objects with repetition and with additional constraints. Consider the integer equation problem of Chapter 1, which asks for the number of nonnegative integer solutions of $x_1 + x_2 + \cdots + x_n = r$, in which we imposed no other restrictions on the n variables. How do we solve this problem if we now restrict each variable x_i to be an element of a set V_i? A typical problem: Find the number of ways to make 62 cents involving quarters, dimes, nickels, and cents. The solution is the number of solutions in nonnegative integers of $q + d + n + c = 62$, where q is in the set $Q = \{0, 25, 50\}$, d is in $D = \{0, 10, 20, 30, 40, 50, 60\}$, n is in $N = \{0, 5, 10, 15, 20, 25, \ldots, 60\}$, and c is in $C = \{0, 1, 2, 3, 4, \ldots, 60, 61, 62\}$. Before we develop a procedure to solve this "change-making problem" using generating functions, let us examine a simpler problem.

Example 2.1.1

Find the number of integer solutions of $a + b + c = 10$, where each variable is at least 2 and at most 4.

Solution (By Explicit Enumeration):

a	b	c
2	4	4
3	4	3
3	3	4
4	2	4
4	4	2
4	3	3

Thus there are six different solutions for this problem.

Now we introduce three polynomials p_a, p_b, and p_c, one for each variable. Since each variable can be 2 or 3 or 4, in this case each polynomial is defined as $x^2 + x^3 + x^4$ and we multiply these three polynomials to obtain a polynomial $p(x)$ involving powers of x with exponents ranging from 6 to 12. This polynomial $p(x)$ is an example of a generating function. Since $a + b + c = 10$ we now look for the coefficient of the tenth power of x in the polynomial $p(x)$. In how many ways can we form the tenth power of x in $p(x)$? For example, we can choose x^2 from p_a, x^4 from p_b, and x^4 from p_c and multiply them. This is just one way of getting the tenth power of x and this corresponds to the solution $a = 2$, $b = 4$, and $c = 4$. In other words, every solution of the problem corresponds to exactly one way of obtaining the tenth power of x in $p(x)$. So the number of solutions of the problem is the coefficient of the tenth power of x in the function $p(x) = (x^2 + x^3 + x^4)^3$. By ordinary polynomial multiplication we see that this coefficient is 6.

DEFINITION 2.1.1

(a) A **power series** is an infinite series of the form $a_0 + a_1x + a_2x^2 + a_3x^3 + \cdots$, where a_i ($i = 0, 1, 2, \ldots$) are real numbers and x is a variable.

(b) If $a_0 + a_1x + a_2x^2 + \cdots$ and $b_0 + b_1x + b_2x^2 + \cdots$ are two power series, then (1) the **sum** of the two power series is a power series in which the coefficient of x^r is $a_r + b_r$ and (2) the **product** of the two power series is a power series in which the coefficient of x^r is $(a_0b_r + a_1b_{r-1} + a_2b_{r-2} + \cdots + a_rb_0)$.

(c) If a_r ($r = 0, 1, 2, \ldots$) is the number of ways of selecting r objects in a certain combinatorial problem (or, more generally, the number of solutions of a combinatorial problem), the **ordinary generating function** for this combinatorial problem is the power series $a_0 + a_1x + a_2x^2 + a_3x^3 + \cdots$.

Any polynomial in x is a power series in x. For example, the polynomial $3x^2 + 2x^4$ can be written as $0 + 0 \cdot x + 3x^2 + 0 \cdot x^3 + 2x^4 + 0 \cdot x^5 + 0 \cdot x^6 + \cdots$. The addition and multiplication procedures in the definition are obvious generalizations of ordinary polynomial addition and multiplication.

Now consider the problem $a + b + c = r$, where a, b, and c are at least 2 and at most 4. Then r varies from 6 to 12. For a fixed choice of r, let a_r be the number of solutions in integers. Then a_r is the coefficient of x^r in the generating function $g(x)$ of the problem where $g(x) = (x^2 + x^3 + x^4)^3$, which is equal to $x^6 + 3x^7 + 6x^8 + 7x^9 + 6x^{10} + 3x^{11} + x^{12}$.

Example 2.1.2

The number of ways of choosing r elements from a set of n elements is $C(n, r)$, and so the generating function for this combinatorial problem is $g(x)$, where

$$g(x) = C(n, 0) + C(n, 1)x + C(n, 2)x^2$$

$$+ \cdots + C(n, r)x^r + \cdots C(n, n)x^n$$

which is the binomial expansion for $(1 + x)^n$.

Example 2.1.3

Find the generating function $g(x)$ in which the coefficient of x^r is a_r, where a_r is the number of solutions in nonnegative integers of the equation $2a + 3b + 5c = r$.

Solution. We write $A = 2a$, $B = 3b$, and $C = 5c$ and seek the number of solutions of $A + B + C = r$, where A is in the set $\{0, 2, 4, 6, \ldots\}$, B is in $\{0, 3, 6, 9, \ldots\}$, and C is in $\{0, 5, 10, 15, \ldots\}$. Thus the generating function is $g(x) = (1 + x^2 + x^4 + x^6 + \cdots)(1 + x^3 + x^6 + x^9 + \cdots)(1 + x^5 + x^{10} + x^{15} + \cdots)$.

Example 2.1.4

The number of solutions in nonnegative integers of $a + b + c = 4$ (with no other constraints on the variables) is the coefficient of x^4 either in $g(x) = (1 + x + x^2 + x^3 + x^4)^3$ or in $h(x) = (1 + x + x^2 + x^3 + x^4 + x^5 + \cdots)^3$. Notice that $g(x)$ is a polynomial in x, whereas $h(x)$ is a power series that is not a polynomial.

Example 2.1.5

If a_r is the number of ways of selecting r marbles from a collection of red, blue, and white marbles such that the number of red marbles selected is at most two, the number of blue marbles selected is at most three and the number of white marbles selected is at most four, then a_r is the coefficient of x^r in the generating function

$$g(x) = (1 + x + x^2)(1 + x + x^2 + x^3)(1 + x + x^2 + x^3 + x^4)$$

Equivalently, the coefficient of x^r in $g(x)$ is the number of solutions in nonnegative integers of $a + b + c = r$, where a is at most 2, b is at most 3, and c is at most 4.

2.2 ORDINARY GENERATING FUNCTIONS

In Example 2.1.1, we saw that the number of computational steps involved in finding the number of solutions by explicit enumeration is exactly equal to the number of computational steps involved in finding the coefficient of the tenth power of x in the generating function, and therefore the generating function method was in no way more efficient than the explicit enumeration method. We shall now develop some simple techniques for calculating the coefficients of generating functions without actually carrying out the polynomial (power series) multiplication procedure.

THEOREM 2.2.1

(a) Let a_r be the coefficient of x^r in $g(x) = (1 + x + x^2 + x^3 + \cdots)^n$. Then $a_r = C(r + n - 1, r)$.

(b) $(1 - x^m)^n = 1 - C(n, 1)x^m + C(n, 2)x^{2m} - \cdots + (-1)^n x^{nm}$.

(c) $(1 + x + x^2 + \cdots + x^{m-1})^n = (1 - x^m)^n (1 + x + x^2 + \cdots)^n$.

Proof:

(a) The function $g(x)$ is the generating function associated with the combinatorial problem that seeks the number a_r of solutions in nonnegative integers of the equation $y_1 + y_2 + \cdots + y_n = r$ and it was proved in Chapter 1 that the number of solutions is $C(r + n - 1, n - 1)$, which is equal to $C(r + n - 1, r)$.

(b) Put $t = (-x^m)$ in the binomial expansion of $(1 + t)^n$.

(c) It is easily verified (in a formal sense) that

$$1 + x + x^2 + \cdots + x^{m-1} = (1 - x^m)(1 + x + x^2 + \cdots)$$

Now take the nth power on both sides of this equation.   ◇

Example 2.2.1

Find the number of solutions in integers of the equation $a + b + c + d = 27$, where each variable is at least 3 and at most 8.

Solution.   The number of solutions is the coefficient of the twenty-seventh power of x in $g(x) = (x^3 + x^4 + \cdots + x^8)^4$, and this number is the coefficient of the fifteenth power of x in $h(x) = (1 + x + \cdots + x^5)^4$. By (c) of Theorem 2.2.1,

$$h(x) = (1 - x^6)^4(1 + x + x^2 + \cdots)^4$$

By (b) of this theorem,

$$(1 - x^6)^4 = 1 - C(4, 1)x^6 + C(4, 2)x^{12} + \cdots$$

and by (a) of the same theorem,

$$(1 + x + x^2 + \cdots)^4 = 1 + C(4, 1)x + C(5, 2)x^2 + C(6, 3)x^3 + \cdots$$

Thus the coefficient of the fifteenth power of x in $h(x)$ is equal to

$$C(18, 15) - C(4, 1)C(12, 9) + C(4, 2)C(6, 3)$$

Example 2.2.2

Find the coefficient of the twenty-fourth power of x in $(x^3 + x^4 + \cdots)^5$.

Solution.   The desired number is the coefficient of the ninth power of x in $g(x) = (1 + x + x^2 + \cdots)^5$, which is equal to $C(13, 4)$.

If $a_0 + a_1x + a_2x^2 + \cdots + a_rx^r + \cdots$ is the power series expansion of a function $g(x)$, then $g(x)$ is the ordinary generating function for the sequence a_r. From a given generating function it is possible to build new generating functions for different choices of a_r, and this is the content of the next theorem, the proof of which is left as an exercise.

THEOREM 2.2.2

If $g(x)$ is the generating function for a_r and $h(x)$ is the generating function for b_r then:

(a) $Ag(x) + Bh(x)$ is the generating function for $Aa_r + Bb_r$.

(b) $(1 - x)g(x)$ is the generating function for $a_r - a_{r-1}$.

(c) $(1 + x + x^2 + \cdots)g(x)$ is the generating function for $(a_0 + a_1 + a_2 + \cdots + a_r)$.

(d) $g(x)h(x)$ is the generating function for $(a_0b_r + a_1b_{r-1} + a_2b_{r-2} + \cdots + a_rb_0)$.

(e) $xg'(x)$ is the generating function for ra_r, where $g'(x)$ is the derivative of $g(x)$ with respect to x.

When the symbol x is a real number with absolute value less than 1 it can be actually verified that $(1 - x)(1 + x + x^2 + x^3 + \cdots) = 1$. (For a proof, see the discussion of the convergence of geometric series in any introductory calculus book. In this book we are more interested in the coefficients of the powers of x considered as a symbol than with issues of convergence.) Thus we write

$$g(x) = 1 + x + x^2 + x^3 + \cdots = \frac{1}{1 - x} = (1 - x)^{-1}$$

$$h(x) = (g(x))^n = \left(\frac{1}{1 - x}\right)^n = (1 - x)^{-n}$$

where $g(x)$ is the generating function for $a_r = 1$ and $h(x)$ is the generating function for $a_r = C(r + n - 1, r)$.

Example 2.2.3

Find the generating function for $a_r = 3r + 5r^2$.

Solution. Let $g(x) = 1/(1 - x)$. The generating function for 1 is $g(x)$. So the generating function for r is $xg'(x)$, by (e) of Theorem 2.2.2. By applying this principle once more we see that the generating function for r^2 is $x(xg'(x))'$. Thus the desired generating function is $3xg'(x) + 5x(xg'(x))'$, which is equal to

$$\frac{3x}{(1 - x)^2} + \frac{5x + 5x^2}{(1 - x)^3}$$

2.3 EXPONENTIAL GENERATING FUNCTIONS

The generating functions we have seen thus far are referred to as "ordinary" generating functions because they were associated with selection problems in which order was not relevant. In other words, they are used to solve combinatorial problems of distribution of identical (undistinguishable) objects into distinct locations. Now we turn to problems of arrangements in which order plays a significant role. For example, the problem of finding the number of ways in which 5 red (undistinguishable) marbles can be put in 3 distinct boxes is a problem in which order is not relevant, whereas the problem of finding the number of ways of arranging 5 marbles in a row using three different types of marbles (red, blue, and white, say) is a problem in which order plays a crucial role. The arrangement RRBBW (red, red, blue, blue, white) is not the same as the arrangement RBRBW, even though both the arrangements use the same number of red, blue, and white marbles. Generating functions that are defined in connection with such combinatorial problems, where order is relevant, are called **exponential generating functions.** Let us analyze this example of arranging marbles before we give a formal definition of exponential generating functions.

Example 2.3.1

Find the number of ways of arranging 5 marbles in a row using marbles of three colors (red, blue, and white) so that in each arrangement there is at least one marble of each color, *assuming that there are at least 3 marbles of each color at our disposal.*

Solution.   Let the number of red, blue, and white marbles in a particular arrangement be r, b, and w. Then $r + b + w = 5$, where each variable is an integer that is at least 1. We know (from our discussion of generalized permutations in Chapter 1) that with this particular choice or r, b, and w there are $(5!)/(r!)(b!)(w!)$ ways of arranging 5 marbles in a row. Thus the total number of arrangements will be the sum of all expressions of the form $((r + b + w)!)/(r!)(b!)(w!)$, where $r + b + w = 5$ and each variable is an integer that is at least 1. The choices of r, b, and w are as follows:

r	b	w
3	1	1
1	3	1
1	1	3
2	2	1
2	1	2
1	2	2

Thus the number of arrangements will be

$$\frac{(5!)}{(3!)(1!)(1!)} + \frac{(5!)}{(1!)(3!)(1!)} + \frac{(5!)}{(1!)(1!)(3!)} + \frac{(5!)}{(2!)(2!)(1!)}$$

$$+ \frac{(5!)}{(2!)(1!)(2!)} + \frac{(5!)}{(1!)(2!)(2!)} = 150$$

Now it can easily be verified that the coefficient of $x^5/5!$ in the function $g(x)$, where $g(x) = (x/(1!) + x^2/(2!) + x^3/(3!))^3$ is precisely the sum of the six expressions obtained in the preceding paragraph giving the total number of arrangements. The function $g(x)$ is an example of an exponential generating function. As in the case of ordinary generating functions, we take the third power of a polynomial (representing the three distinct colors), and the powers of the variable in the polynomial are 1, 2, and 3, indicating that the number of ways a marble of a particular color can appear in an arrangement is 1, 2, or 3. The significant difference here is that unlike the ordinary generating function, the coefficient of the x^r in the polynomial is $1/(r!)$ and the solution of the combinatorial problem is the coefficient of $x^r/(r!)$ in the exponential generating function. Is there an easier method in this problem to find the coefficient of $x^5/(5!)$ in $g(x)$? Let $h(x) = (e^x - 1)^3$, where e^x is the exponential function (power series) defined by $1 + x + x^2/(2!) + x^3/(3!) + \cdots$, where x is any real variable. Then the required coefficient is that of $x^5/(5!)$ in $h(x) = (e^{3x} - 3e^{2x} + 3e^x - 1)$ and this coefficient is $3^5 - (3)2^5 + 3 = 150$.

DEFINITION 2.3.1

If b_r $(r = 0, 1, 2, \ldots)$ is the solution of a combinatorial problem, the power series $g(x)$ defined by $b_0 + b_1x + (b_2x^2)/(2!) + (b_3x^3)/(3!) + \cdots$ is called the **exponential generating function** for that problem.

Example 2.3.2

Find the exponential generating function for b_r, the number of ways of arranging r distinct elements from a set of n elements.

Solution.   Of course, $b_r = P(n, r)$, so the exponential generating function for this problem is $g(x)$, which is a power series in which the coefficient of $(x^r) = [P(n, r)]/(r!) = C(n, r)$. Thus the exponential generating function for $P(n, r)$ is $(1 + x)^n$, which is the same as the ordinary generating function of $C(n, r)$.

Example 2.3.3

Find the number of ways of arranging 5 marbles in a row using marbles of three colors (red, blue, and white) so that each arrangement has at least one marble of each color assuming that we have at most 3 red, at most 2 white, and at most 2 blue marbles at our disposal. (Notice the difference between this example and Example 2.3.1.)

Solution.   In this case the number of arrangements will be the coefficient of $x^5/5!$ in $g(x) = (x + x^2/2! + x^3/3!)(x + x^2/2!)^2$ and this coefficient is the same as the coefficient of $x^5/5!$ in

$$h(x) = \left(x + \frac{x^2}{2!} + \frac{x^3}{3!} + \cdots\right)\left(x + \frac{x^2}{2!}\right)^2$$

$$= (e^x - 1)\left(x + \frac{x^2}{2!}\right)^2$$

The analysis in Examples 2.3.1 and 2.3.3 generalizes as follows:

THEOREM 2.3.1

Suppose that there are k types of objects.

(a) If there is an unlimited supply of objects in each of these types, then the number of r-permutations ($r = 1, 2, \ldots$) using objects from these k types is the coefficient of $x^r/r!$ in the exponential generating function

$$g(x) = \left(1 + x + \frac{x^2}{2!} + \frac{x^3}{3!} + \cdots\right)^k = e^{kx}$$

(b) If the supply of objects in type i is at most n_i (where $i = 1, 2, \ldots, k$), the number of r-permutations will be the coefficient of $x^r/r!$ in

$$h(x) = \left(1 + x + \frac{x^2}{2!} + \cdots + \frac{x^{n_1}}{n_1!}\right)\left(1 + x + \frac{x^2}{2!}\right.$$

$$\left. + \cdots + \frac{x^{n_2}}{n_2!}\right) \cdots \cdot \left(1 + x + \frac{x^2}{2!} + \cdots + \frac{x^{n_k}}{n_k!}\right)$$

(c) $(n!) \cdot S(r, n) = $ coefficient of $x^r/r!$ in $(e^x - 1)^n$ where $S(r, n)$ is a Stirling number of the second kind defined in Chapter 1.

Example 2.3.4

(a) Find the number of r-permutations that can be formed using the letters I, M, S, and P, where r is a positive integer.

(b) Find the number of r-permutations that can be formed using the letters that appear in the word MISSISSIPPI so that the number of times a letter appears in a permutation is at most equal to the number of times the letter appears in the word.

Solution

(a) The number of r-permutations is the coefficient of $x^r/r!$ in the exponential generating function $g(x) = e^{4x}$, and this coefficient is 4^r.

(b) In the word, the frequencies of the letters are 4, 1, 2, and 4. Thus the number of r-permutations is the coefficient of $x^r/r!$ in

$$h(x) = \left(1 + x + \frac{x^2}{2!} + \frac{x^3}{3!} + \frac{x^4}{4!}\right)(1 + x)$$

$$\cdot \left(1 + x + \frac{x^2}{2!}\right)\left(1 + x + \frac{x^2}{2!} + \frac{x^3}{3!} + \frac{x^4}{4!}\right)$$

where r is at most 11, the sum of the frequencies.

Example 2.3.5

Find the number of ways of accommodating 9 people in 4 rooms such that no room is left unoccupied.

Solution.   If x denotes the number of people assigned to a room, then x is at least one and at most 6, and there are 4 rooms. Thus the exponential generating function for this combinatorial problem is

$$g(x) = \left(x + \frac{x^2}{2!} + \frac{x^3}{3!} + \cdots + \frac{x^6}{6!}\right)^4$$

Now the number of ways of accommodating 9 people in 4 rooms is the coefficient of $(x^9)/(9!)$ in $g(x)$ and this coefficient is equal to the coefficient of $(x^9)/(9!)$ in

$$h(x) = (e^x - 1)^4 = e^{4x} - 4e^{3x} + 6e^{2x} - 4e^x + 1$$

Thus the number of arrangements will be $4^9 - (4)3^9 + (6)2^9 - 4$. [Notice that the number of arrangements is equal to $(4!)S(9, 4)$ where $S(9, 4)$ is the Stirling number of the second kind defined in Chapter 1.]

2.4 NOTES AND REFERENCES

The first comprehensive treatment of generating functions was given by Pierre Simon Marquis de Laplace (1749–1827). But the method of solving problems using generating functions has its origin in the works of Abraham de Moivre (1667–1754). Both Leonhard Euler (1707–1783) and Nicholas Bernoulli (1687–1759) also used this technique in their investigation of certain combinatorial problems: Euler was interested, among other things, in partition problems and Bernoulli was interested in derangement problems. For a thorough treatment of generating functions, see the books on combinatorics by MacMahon (1960) or Riordan (1958). See also Riordan (1964). Other general references include the relevant chapters in the books by Cohen (1978), Krishnamurthy (1986), Liu (1968), Liu (1985), Roberts (1984), Tucker (1984), and Townsend (1987).

2.5 EXERCISES

2.1. Find the ordinary generating functions for the following sequences.
  (a) $\{1, 1, 1, 1, 0, 0, \ldots\}$ (b) $\{0, 0, 0, 0, 1, 1, \ldots\}$
  (c) $\{1, 1, 1, 1, \ldots\}$ (d) $(1, -1, 1, -1, \ldots)$

2.2. Find the ordinary generating functions for the following sequences.
  (a) $\{1, 2, 3, 4, \ldots\}$ (b) $\{1, -2, 3, -4, \ldots\}$

2.3. Find the sequence corresponding to the following ordinary generating functions.
  (a) $(2 + x)^4$ (b) $x^2 + e^x$ (c) $x^3(1 - x)^{-1}$

2.4. Find the coefficient of x^7 in $(1 - x)^k$ when $k = 9$ and $k = -9$.

2.5. Find the coefficient of x^7 in $(1 + x)^k$ when $k = 9$ and $k = -9$.

2.6. Find the coefficient of x^{23} in $(x^3 + x^4 + \cdots)^5$.

2.7. Find the ordinary generating function $f(x)$ that can be associated with the combinatorial problem of finding the number of solutions in positive integers of the equation $a + b + c + d = r$.

2.8. Find the ordinary generating function associated with the problem of finding the number of solutions in nonnegative integers of the equation $3a + 2b + 4c + 2d = r$.

2.9. Find the number of solutions in integers of the equation $p + q + r + s = 27$ where each variable is at least 3 and at most 8.

2.10. Find the number of solutions of $x_1 + x_2 + \cdots + x_n = r$ where each variable is either 0 or 1.

2.11. If three distinct dice (marked A, B, and C) are thrown, find the number of ways of getting a total of 13.

2.12. Solve Problem 2.11 if the first die (marked A) shows an even number.

2.13. Find the number of ways of allocating 9 identical objects to 3 different locations (numbered as first, second, and third) such that each location gets at least one object and the third location does not get more than 3 objects.

2.14. Find the ordinary generating function associated with the combinatorial problem of choosing 9 marbles from a bag that has 3 indentical red marbles, 4 identical blue marbles, and 5 identical green marbles such that in every choice all colors are represented and no color has absolute majority.

2.15. Prove: $(1 + x^m)^n = 1 + C(n, 1)x^m + C(n, 2)(x^m)^2 + \cdots + (x^m)^n$.

2.16. Find the number of solutions in integers of the equation $a + b + c + d + e + f = 20$, where a is at least 1 and at most 5 and the other variables are at least 2 by (**a**) the method developed in Chapter 1, and (**b**) considering a suitable generating function.

2.17. There are 10 identical gift boxes. Each box has to be wrapped with either red or blue or green or yellow wrapping paper. The available red paper can be used to wrap at most 2 boxes and the available blue paper can be used to wrap at most 3 boxes. Write down the ordinary generating function associated with the problem of finding the number of ways of wrapping these 10 boxes.

2.18. There are 9 people in a group. Find the number of ways of collecting \$9.00 from this group if the leader of the group will give at least \$1.00 and at most \$2.00 and every other member will give at most \$1.00.

2.19. Find the ordinary generating function associated with the problem of finding the number of solutions in integers of the inequality $a + b + c \leq r$, where each variable is at least 2 and at most 5.

2.20. A participant in a contest is rated on a scale of 1 to 6 by each of the 4 judges. To be a finalist a participant has to score at least 22. Find the number of ways the judges can rate a participant so that she can be a finalist.

2.21. An Antarctic expedition group consists of scientists representing the United States, the USSR, and England. Find the number of ways of forming a group of 9 scientists so that none of these three countries has an absolute majority in the group.

2.22. Use a combinatorial argument to prove that the coefficient of x^{2n+1} in $f(x)$ is equal to the coefficient of x^{2n-2} in $g(x)$, where $f(x) = (1 + x + \cdots + x^n)^3$ and $g(x) = (1 + x + x^2 + \cdots + x^{n-1})^3$. Find this coefficient.

2.23. Find the number of ways of distributing 8 apples and 6 oranges to 3 children so that each child can get at least 2 apples and at most 2 oranges.

2.24. Prove **(a)** $1 + 2 + 3 + \cdots + r = [r(r + 1)]/2$ and **(b)** $1 + 2^2 + 3^2 + \cdots + r^2 = [r(r + 1)(2r + 1)]/6$.

2.25. Find the number of ways of storing p identical red marbles in m boxes in one shelf and q identical blue marbles in n boxes in another shelf so that no box is empty. (Since no box will be empty p cannot be less than m and q cannot be less than r.) Solve the problem when $p = 6$, $q = 7$, $m = 3$, and $n = 4$.

2.26. Find the ordinary generating functions for the sequences:
  (a) $\{a_r\}$ where $a_r = k^r$, where k is a constant
  (b) $\{b_r\}$ where $b_r = rk^r$
  (c) $\{c_r\}$ where $c_r = k + 2k^2 + 3k^3 + \cdots + rk^r$

2.27. The sum of four positive integers in nondecreasing order is r and a_r is the number of ways of choosing these four integers. Find the ordinary generating function associated with the sequence $\{a_r\}$.

2.28. If X is a set with n elements, show that the number of subsets of X with $(r - 1)$ elements is equal to the number of solutions of the equation $y_1 + y_2 + \cdots + y_r = (n - 1)$, where the first two variables are nonnegative and the other variables are positive.

2.29. Let $X = \{1, 2, 3, \ldots, n\}$. Find the number of subsets of X such that each subset has r elements and no two elements in a subset are consecutive integers.

2.30. If r is a positive integer, a **partition** of r is a collection of positive integers whose sum is r. A partition is distinct if the integers in it are distinct. For example $\{3, 1\}$ is a distinct partition of 4, whereas $\{2, 2\}$ is a partition of 4 that is not distinct. The number of partitions of r is denoted by $p(r)$ and the number of distinct partitions of r is denoted by $p_d(r)$. Obtain the generating functions to compute $p(r)$ and $p_d(r)$.

2.31. Show that the number of distinct partitions of a positive integer r is the same as the number of partitions of r into odd positive integers.

2.32. Let $p(r; n)$ be the number of partitions of r such that in each partition no element exceeds n. Show that $p(r; r) = p(r)$.

2.33. Show that every nonnegative integer can be written uniquely in binary form.

2.34. Find the exponential generating function associated with the following sequences.
  (a) $\{1, 1, 1, 1, 0, 0, 0, 0, \ldots\}$ **(b)** $\{0, 0, 0, 0, 1, 1, 1, 1, \ldots\}$
  (c) $\{1, 2, 2^2, 2^3, 2^4, \ldots\}$ **(d)** $\{1, 1, 2 \cdot 2, 3 \cdot 2^2, 4 \cdot 2^3, \ldots\}$

2.35. **(a)** Use a combinatorial argument to prove that $(e^x)^n = e^{nx}$.
  (b) Prove that $e^x + e^{-x} = 2(1 + x^2/2! + x^4/4! + \cdots)$.
  (c) Prove that $e^x - e^{-x} = 2(x + x^3/3! + x^5/5! + \cdots)$.

2.36. Let $X = \{A, B, C, D\}$. Using exponential generating functions, obtain **(a)** the number of r-permutations that can be formed using these four letters such that in each permutation there is at least one A, at least one B, and at least one C, and **(b)** the number of r-permutations that can be formed such that in each permutation there is an even number of A's and an odd number of B's.

2.37. Find the number of r-digit binary numbers that can be formed using an even number of 0's and an even number of 1's.

2.38. (a) Find the number of ways the headquarters of a company can allocate nine new identical computers to four distinct branch offices so that each office gets at least one new computer.
(b) Find the number of ways the headquarters of a company can allocate nine new employees to four distinct branch offices so that each office gets at least one new employee.

2.39. Find the number of (a) permutations of the letters which appear in the word MISSISSIPPI, (b) the number of 6-permutations of letters that appear in this word, and (c) the number of 6-permutations of letters from this word such that in each permutation every letter of the word appears at least once.

2.40. Find the number nine-digit sequences that can be formed using the digits 0, 1, 2, and 3 such that (a) each sequence has an even number of 0's, (b) each sequence has an odd number of 0's, (c) each sequence has an even number of 0's and an odd number of 1's, (d) the total number of 0's and 1's is odd, and (e) no digit appears exactly twice.

2.41. Obtain the appropriate generating function associated with the combinatorial problem of finding the number of codewords of length r from an alphabet consisting of five distinct letters such that in each codeword every letter of the alphabet appears at least once and the first letter appears an even number of times.

CHAPTER

3

Recurrence Relations

3.1 INTRODUCTION

Consider a sequence a_0, a_1, a_2, . . . , where a_r is the solution of a certain combinatorial problem that depends on the input r. In Chapter 2 we discussed some methods to compute a_r using generating functions. In some cases it will be possible to reduce the computation of the rth term of the sequence to earlier members of the sequence if a_r can be expressed as a function of the earlier elements of the sequence. For example, consider the arithmetic progression sequence 4, 7, 10, 13, 16, . . . , where the initial number a_0 is 4 and the common difference d is 3. Then the rth term of the sequence can be expressed in terms of the $(r - 1)$th term by the equation $a_r = a_{r-1} + d$. This equation is an **example of a recurrence relation.** The condition $a_0 = 4$ is called the **initial condition** of this relation. Obviously, once the initial condition and the common difference are known any arbitrary term can be obtained by computing a_1, a_2, . . . , sequentially. Or we can obtain the rth term by **solving** the **recurrence**

relation. In this case the **solution** is $a_r = 4 + 3r$, where r is any nonnegative integer. Similarly, if we take the geometric progression sequence 4, 4.3, 4.3^2, 4.3^3, . . . , the recurrence relation is $a_r = 3 \cdot a_{r-1}$ with initial condition $a_0 = 4$ and the solution is $a_r = 4 \cdot 3^r$.

Recurrence Relations and Difference Equations

The **first difference** $d(a_n)$ of a sequence $\{a_n\}$ of real numbers is the difference $a_n - a_{n-1}$. The **second difference** $d^2(a_n)$ is $d(a_n) - d(a_{n-1})$, which is equal to $a_n - 2a_{n-1} + a_{n-2}$. More generally the *k*th **difference** $d^k(a_n)$ is $d^{k-1}(a_n) - d^{k-1}(a_{n-1})$. A **difference equation** is an equation involving a_n and its differences. For example, $3d^2(a_n) + 2d(a_n) + 7a_n = 0$ is a second-order homogeneous difference equation. Observe that every a_i ($i = 0, 1, 2, \ldots, n - 1$) can be expressed in terms of a_n and these differences because $a_{n-1} = a_n - d(a_n)$;

$$a_{n-2} = a_{n-1} - d(a_{n-1}) = [a_n - d(a_n)] - d(a_n - d(a_n))$$

$$= a_n - 2d(a_n) + d^2(a_n)$$

and so on. Thus every recurrence relation can be formulated as a difference equation. On the other hand, using the definition of these differences, any difference equation can be formulated as a recurrence relation. For instance, the difference equation $3d^2(a_n) + 2d(a_n) + 7a_n = 0$ can be expressed as the recurrence relation $12a_n = 8a_{n-1} - 3a_{n-2}$. Thus some authors use the terms *difference equation* and *recurrence relation* interchangeably. The methods for solving recurrence relations were developed originally using the techniques used for solving difference equations. Difference equations are commonly used to approximate differential equations when solving differential equations using computers.

Notice that in a relation of the type $a_r = a_{r-1} + a_{r-2}$ we need to know both a_0 and a_1 to obtain any a_r ($r > 1$), and therefore we need two initial conditions to solve this equation. Thus with the information available in the set of initial conditions of a given recurrence relation in most cases one should be able to compute sequentially any arbitrary term of the sequence. We shall study techniques for solving certain types of recurrence relations later in this section. There are no general methods for solving all recurrence relations.

Example 3.1.1

The recurrence relation $a_r = ra_{r-1}$ with the initial condition $a_0 = 1$ has the solution $a_r = r!$ ($r = 1, 2, \ldots$).

Example 3.1.2

Find a recurrence relation to obtain a_n, the number of ways of arranging n distinct elements in a row.

Solution. There are n ways of choosing an element to be placed in the first position of the row. After placing an element in the first position the number of ways of arranging the remaining $(n - 1)$ elements is a_{n-1}. Thus we have the recurrence relation $a_n = na_{n-1}$ with the initial condition $a_1 = 1$, the solution of which (by the previous example) is $a_n = n!$

Example 3.1.3

Suppose that the interest rate offered by a bank to the depositors is $r\%$ per year. If a_n is the amount on deposit at the end of n years, obtain a recurrence relation for a_n if (a) the interest is simple, and (b) the interest is compounded annually.

Solution. (a) If a_0 is the initial deposit, at the end of year k, ra_0 is added to a_k if the interest is simple. Thus the recurrence relation is $a_{k+1} = a_k + ra_0$, where $k = 0, 1, 2, \ldots$. By iteration we see that the solution is

$$a_{k+1} = a_{k-1} + ra_0 + ra_0 = a_{k-2} + 3ra_0 = \cdots = a_0 + (k + 1)ra_0$$

Thus the solution to the recurrence relation is $a_n = (1 + nr)a_0$.

(b) If the interest is compounded annually, the recurrence relation is $a_{k+1} = a_k + ra_k = (1 + r)a_k$, the solution of which, by iteration, is $a_n = (1 + r)^n a_0$.

Example 3.1.4

Find the recurrence relation for the Fibonacci sequence 1, 2, 3, 5, 8, 13, ... in which the rth term is the sum of the $(r - 1)$th term and the $(r - 2)$th term. Obviously, the relation is $a_n - a_{n-1} - a_{n-2} = 0$ with initial conditions $a_1 = 1$ and $a_2 = 2$. (The numbers that appear in the sequence are called *Fibonacci numbers* and they arise in many areas of combinatorial mathematics. We discuss a solution technique later.)

Recursion and Recurrence

A recurrence relation as we see here is a recursive formula (see Section 3 of Chapter 0) to compute the number of ways to do a procedure involving n objects in terms of the number of ways to do it with fewer objects. This recursive reasoning involved in building a recurrence relation model of a counting problem is the same logic used in designing recursive computer subroutines that call themselves. The basic idea of any recursive procedure in computer science is that it calls itself to solve a problem by solving similar problems that are smaller than the original problem. An important feature of recursion is the concept of working backward. However, we cannot use a recursive program to compute the values in a recurrence relation because recurrence relations are meant for forward tabulation of values, not for backward recursive computation.

3.2 HOMOGENEOUS RECURRENCE RELATIONS

As mentioned before there is no general method of solution for an arbitrary recurrence relation. In what follows we study a broad class of recurrence relations for which solution techniques are known.

DEFINITION 3.2.1

If c_i ($i = 1, 2, \ldots, r$) are constants, a recurrence relation of the form $a_n = c_1 a_{n-1} + c_2 a_{n-2} + \cdots + c_r a_{n-r} + f(n)$ is called a **linear recurrence relation with constant coefficients of order** r. The recurrence relation is **homogeneous** if the function $f(n) = 0$. If $g(n)$ is a function such that $a_n = g(n)$ for $n = 0, 1, 2, \ldots$, then $g(n)$ is a **solution** of the recurrence relation.

Example 3.2.1

It can be verified by substitution that $g(n) = A \cdot 2^n + B \cdot n \cdot 2^n + n^2 \cdot 2^{n-1}$ (where A and B are arbitrary constants) is a solution of the following second-order inhomogeneous linear recurrence equation with constant coefficients: $a_n = 4a_{n-1} - 4a_{n-2} + 2^n$. (We shall study solution techniques to obtain such general solutions in this section.)

THEOREM 3.2.1   (The Principle of Superposition)

If $g_i(n)$, where $i = 1, 2, \ldots, k$, are solutions of

$$a_n = c_1 a_{n-1} + c_2 a_{n-2} + \cdots + c_r a_{n-r} + f_i(n)$$

then any linear combination of the k solutions of the form $A_1 g_1(n) + A_2 g_2(n) + \cdots + A_k g_k(n)$ is a solution of the recurrence relation

$$a_n = c_1 a_{n-1} + c_2 a_{n-2} + \cdots + c_r a_{n-r} + A_1 f_1(n) + \cdots + A_k f_k(n)$$

where A_i $(i = 1, 2, \ldots, k)$ are real numbers. In particular, any linear combination of the solutions of a homogeneous recurrence relation is again a solution of the homogeneous recurrence relation.

Proof:

Let

$$h(n) = A_1 g_1(n) + A_2 g_2(n) + \cdots + A_k g_k(n)$$

Since $g_i(n)$ is a solution of

$$a_n = c_1 a_{n-1} + c_2 a_{n-2} + \cdots + c_r a_{n-r} + f_i(n)$$

we have

$$g_i(n) = c_1 g_i(n - 1) + c_2 g_i(n - 2) + \cdots + c_r g_i(n - r) + f_i(n)$$

and therefore

$$h(n) = c_1 h(n - 1) + c_2 h(n - 2) + \cdots$$
$$+ c_r h(n - r) + A_1 f_1(n) + \cdots + A_k f_k(n)$$

which proves our assertion. $\diamondsuit$

There is a simple technique for solving homogeneous linear recurrence relations with constant coefficients. Let $a_r = x^r$ be a solution of the relation

$$a_n = c_1 a_{n-1} + c_2 a_{n-2} + \cdots + c_r a_{n-r}$$

Then

$$x^n = c_1 x^{n-1} + c_2 x^{n-2} + \cdots + c_r x^{n-r}$$

If we ignore the trivial solution $x = 0$ we get the polynomial equation

$$x^r - c_1 x^{r-1} - c_2 x^{r-2} - \cdots - c_r = 0$$

This polynomial equation of degree r is called the **characteristic equation** of the recurrence relations which has r roots in general. It is quite possible that the equation has multiple roots or some roots are complex.

If x_i $(i = 1, 2, \ldots, r)$ are the r roots of the characteristic equation then

$a_n = (x_i)^n$ is obviously a solution of the homogeneous recurrence relation and therefore by the previous proposition any linear combination of such solutions is also a solution. For example, the second-order homogeneous recurrence relation $a_n = 5a_{n-1} - 6a_{n-2}$ has the characteristic equation $x^2 - 5x + 6 = 0$ the roots of which are $x_1 = 2$ and $x_2 = 3$. Thus $a_n = A(2)^n + B(3)^n$, for any choice of the arbitrary constants A and B, is also a solution of the recurrence relation. Conversely, if all the r roots x_i ($i = 1, 2, \ldots, r$) are real and distinct, it can be proved that every general solution is a linear combination of these solutions $(x_i)^n$.

The r arbitrary constants that appear in the general solution can be evaluated in some cases giving a complete solution (not necessarily unique) to the problem if r initial conditions are known. Existence and uniqueness of the solution are assured if r consecutive initial conditions are known. We state these results as a proposition the proof of which is omitted here.

THEOREM 3.2.2

If the r roots x_i ($i = 1, 2, \ldots, r$) of the characteristic equation of an rth-order linear homogeneous recurrence relation are real and distinct, every general solution of the recurrence relation is a linear combination of the solutions $(x_i)^n$. Moreover, if r consecutive initial values $a_k, a_{k+1}, \ldots, a_{k+r-1}$ of the recurrence relation are known, a solution can be obtained by evaluating the r arbitrary constants using these r consecutive initial values and this solution is unique.

Example 3.2.2

Solve $a_n - 9a_{n-2} = 0$, where:
  (a) $a_0 = 6$, $a_1 = 12$
  (b) $a_3 = 324$, $a_4 = 486$
  (c) $a_0 = 6$, $a_2 = 54$
  (d) $a_0 = 6$, $a_2 = 10$

Solution.   The roots of the characteristic equation $r^2 - 9 = 0$ are 3 and -3. Thus any general solution of the given recurrence relation is of the form $a_n = A(3)^n + B(-3)^n$, where A and B are arbitrary constants to be evaluated using the two given initial conditions.
    (a) If $a_0 = 6$, then $A + B = 6$. If $a_1 = 12$, then $3A - 3B = 12$. Solving these two simultaneous equations in A and B we get $A = 5$ and $B = 1$, giving the unique (because the initial conditions are consecutive) solution to the problem $a_n = 5(3)^n + (-3)^n$. Putting $n = 2, 3, 4, \ldots$, we can compute $a_2 = 54$, $a_3 = 324$, and so on.
    (b) If $a_3 = 324$, then $27A - 27B = 324$. If $a_4 = 486$, then

$81A + 81B = 486$. Solving these two equations, we get $A = 9$ and $B = -3$, once again giving a unique solution.

(c) This time the two given initial conditions are not consecutive. $a_0 = 6$ implies that $A + B = 6$ and $a_2 = 54$ implies that $9A + 9B = 54$, giving one equation $A + B = 6$ to determine the two constants. For example, $A = 2$, $B = 4$ gives $a_n = 2(3)^n + 4(-3)^n$, which defines $a_0 = 6$, $a_1 = -6$, $a_2 = 54$, $a_3 = -54$, and so on. But $A = 1$, $B = 5$ gives $a_n = (3)^n + 5(-3)^n$, which defines $a_0 = 6$, $a_1 = -12$, $a_2 = 54$, $a_3 = -108$. Thus the solution is not unique.

(d) The nonconsecutive initial conditions imply that $A + B = 6$ and $9A + 9B = 10$, showing that there is no solution.

Example 3.2.3

Solve the Fibonacci recurrence relation $a_n = a_{n-1} + a_{n-2}$ with the consecutive initial conditions $a_0 = 1$ and $a_1 = 1$.

Solution.   The characteristic equation is $x^2 - x - 1 = 0$, the two roots of which are $x_1 = (1 + \sqrt{5})/2$ and $x_2 = (1 - \sqrt{5})/2$. Thus every solution is of the form $a_n = A(x_1)^n + B(x_2)^n$, where A and B are arbitrary constants. Using the initial conditions we get $A = (5 + \sqrt{5})/10$ and $B = (5 - \sqrt{5})/10$. On substituting these values of the arbitrary constants in the solution, we get the unique solution $a_n = [t^{n+1} - (1 - t)^{n+1}]/\sqrt{5}$, where t is the irrational number $(\sqrt{5} + 1)/2$, known as the *golden ratio*.

Example 3.2.4

Find the number of subsets of a set that has n elements.

Solution.   Let the set be $X = \{1, 2, 3, \ldots, n\}$ and suppose that the total number of subsets of X is a_n. Now every subset of X belongs to one of the following two classes A and B: (a) n is not an element in any of the subsets in class A and (b) n is an element in every subset in class B.

By our assumption class A has a_{n-1} subsets of X. The only way we can get a subset of X that contains n is by adjoining n to a subset from class A. So the number of subsets in class B is exactly equal to the number of subsets in class A. In other words, we have the recurrence relation $a_n = 2a_{n-1}$ with the initial condition $a_0 = 1$. Recall that the empty set is a subset of every set. The characteristic equation is $x - 2 = 0$, giving the unique solution $a_n = 2^n$.

Example 3.2.5

Find the unique solution of the recurrence relation $a_n = 3a_{n-1} + 4a_{n-2} - 12a_{n-3}$, where $a_0 = 2$, $a_1 = 5$, and $a_2 = 13$.

Solution. The roots of the characteristic equation $x^3 - 3x^2 - 4x + 12 = 0$ are 2, -2, and 3. So the general solution is $a_n = p(2)^n + q(-2)^n + r(3)^n$ and the initial conditions imply that

$$p + q + r = 2$$

$$2p - 2q + 3r = 5$$

$$4p + 4q + 9r = 13$$

On solving this linear system, we get $p = 1$, $q = 0$, and $r = 1$. Thus the unique solution is $a_n = 2^n + 3^n$.

We next examine the case when the characteristic equation of a recurrence relation has **repeated (multiple) roots.** Consider, for example, the recurrence relation $a_n = 4a_{n-1} - 4a_{n-2}$, which has the characteristic equation $(x - 2)^2 = 0$ with roots $x_1 = 2$ and $x_2 = 2$. In other words, 2 is a repeated root of multiplicity 2. Of course, $A(2)^n$ is a solution of the recurrence relation, but every solution need not be of this form. It can easily be verified that $B \cdot n \cdot (2)^n$ is also a solution, so by the principle of superposition $A(2)^n + B \cdot n \cdot (2)^n$ is also a solution for any A and B. It turns out that every general solution of the recurrence relation is of this form. The fact that $A(2)^n$ by itself cannot be a general solution is also obvious because if we consider two consecutive initial conditions, say $a_0 = 1$ and $a_1 = 4$, we see that $A = 1$ and also $A = 2$.

When the characteristic equation has multiple roots, the following result is a generalization of the previous theorem. The proof is omitted here. For details, see Roberts (1984).

THEOREM 3.2.3

(a) Suppose that $(x - t)^s$ is a factor of the characteristic equation so that t is a root of multiplicity s. Then

$$u = (t)^n(A_1 + A_2n + A_3n^2 + \cdots + A_sn^{s-1})$$

is a solution of the recurrence relation where A_j ($j = 1, 2, \ldots, s$) are arbitrary constants. This solution u is called a **basic solution** of the relation with respect to the root t.

(b) Suppose that the roots of the recurrence relation are t_k, where the multiplicity of t_k is s_k ($k = 1, 2, \ldots, q$), and suppose that u_k is a basic solution with respect to the root t_k. Then every solution of the recurrence relation is the sum of these q basic solutions.

Example 3.2.6

Find the general solution of the recurrence relation the roots of which are 2, 2, 2, -3, 4, and 4.

Solution. The basic solution for the repeated root 2 is $u_1 = 2^n(A_1 + A_2n + A_3n^2)$. The basic solution for the root -3 is $u_2 = A_4(-3)^n$. The basic solution for the repeated root 4 is $u_3 = 4^n(A_5 + A_6n)$. Thus the general solution is $a_n = u_1 + u_2 + u_3$.

3.3 INHOMOGENEOUS RECURRENCE RELATIONS

We now proceed to analyze linear recurrence relations with constant coefficients of the type $a_n = h_n + f(n)$, where $h_n = c_1a_{n-1} + c_2a_{n-2} + \cdots + c_ra_{n-r}$ and $f(n)$ is a function of n. Here the relation $a_n = h_n$ is called the **homogeneous part** of the given inhomogeneous relation. If $a_n = u_n$ is a solution of the homogeneous part and if $a_n = v_n$ is any solution of the given inhomogeneous relation, then by the principle of superposition we know that $a_n = u_n + v_n$ is also a solution of the same inhomogeneous relation. If u_n has r arbitrary constants, then $u_n + v_n$ also has r arbitrary constants. If r consecutive initial conditions of the inhomogeneous relation are known, these initial conditions can be used to define a linear system of r equations in r variables giving a unique solution. In other words, *if u_n is a general solution of the homogeneous part of an inhomogeneous recurrence relation, and if v_n is a particular solution of the inhomogeneous recurrence relation, then $u_n + v_n$ is a general solution of the same inhomogeneous relation.*

Example 3.3.1

Find the general solution of $a_n = 5a_{n-1} - 6a_{n-2} + 6(4)^n$.

Solution.   The solution of the homogeneous part is $u_n = A(2)^n + B(3)^n$, where A and B are arbitrary constants. It can be verified that $v_n = (48)(4)^n$ is a particular solution of the given inhomogeneous relation. Therefore, the general solution of the relation is $a_n = u_n + v_n$.

Unlike the homogeneous case there is no general method to obtain a particular solution for an arbitrary inhomogeneous problem. However, there are techniques available for certain special cases. We have two such special cases: (1) $f(n) = n^k$, where k is a nonnegative integer, and (2) $f(n) = (q)^n$, where q is a rational number not equal to 1. The principle of superposition is called for if there is a linear combination of functions of these two types. These techniques are as follows:

1. If $f(n) = c(q)^n$ (where c is a known constant) and if q is not a root of the characteristic equation, then the choice for the particular solution is $A(q)^n$,

where A is a constant that is to be evaluated by substituting $a_n = A(q)^n$ in the inhomogeneous equation. If q is a root of the characteristic equation with multiplicity k, the choice for the particular solution is $A(n)^k(q)^n$.

2. If $f(n) = c(n)^k$ and if 1 is not a root of the characteristic equation, a polynomial in n of degree k of the form $A_0 + A_1 n + A_2 n^2 + \cdots + A_k n^k$ is the choice for the particular solution. If 1 is a root of multiplicity t, $A_0 n^t + A_1 n^{t+1} + \cdots + A_k n^{t+k}$ is the choice.

Example 3.3.2

If the characteristic equation of a certain inhomogeneous recurrence relation is $(x - 1)^2(x - 2)(x - 3)^2 = 0$, find the choice for the particular solution when

    (a) $f(n) = 4n^3 + 5n$
    (b) $f(n) = 4^n$
    (c) $f(n) = 3^n$

Solution. The roots of the characteristic equation are 1 (with multiplicity 2), 2 (with multiplicity 1), and 3 (with multiplicity 2). Let u_n be the general solution of the homogeneous part and v_n be the choice of a particular solution. Then

$$u_n = c_1 + c_2 n + c_3 2^n + c_4 3^n + c_5 \cdot n \cdot 3^n$$

    (a) $v_n = An^2 + Bn^3 + Cn^4 + Dn^5$
    (b) $v_n = A \cdot 4^n$
    (c) $v_n = A \cdot n^2 \cdot 3^n$

Example 3.3.3

Discuss the solution of $a_n = ka_{n-1} + f(n)$, where k is a constant.

Solution. As before, $a_n = u_n + v_n$, where u_n is the solution of the homogeneous part and v_n is a particular solution.

    Case (1): $k = 1$. $u_n = c$, where c is an arbitrary constant, so $a_n = c + v_n$, where the nature of v_n depends on $f(n)$ and the fact that u_n is a constant. However,

$$a_n = a_{n-1} + f(n)$$
$$a_{n-1} = a_{n-2} + f(n-1)$$
$$\vdots$$
$$a_2 = a_1 + f(2)$$
$$a_1 = a_0 + f(1)$$

Adding these n equations, we have

$$a_n = a_0 + f(1) + f(2) + \cdots + f(n)$$

Thus

$$f(1) + f(2) + \cdots + f(n) = a_n - a_0 = c + v_n - a_0$$

where the value of c can be evaluated using the initial condition a_0.

Case (2): k not equal to 1. $u_n = c \cdot k^n$, and as before, v_n depends on $f(n)$ and u_n.

Example 3.3.4

Evaluate the sum of the squares of the first n positive integers.

Solution.   Write $f(n) = n^2$. We have to compute $f(1) + f(2) + \cdots + f(n)$. Consider the recurrence relation $a_n = a_{n-1} + n^2$ with $a_0 = 0$. The homogeneous part gives the solution $u_n = c$. The choice for the particular solution is $v_n = An + Bn^2 + Cn^3$. Thus $a_n = c + An + Bn^2 + Cn^3$. The initial condition implies that $c = 0$. Substituting for a_n in the recurrence relation, we get

$$An + Bn^2 + Cn^3 = A(n - 1) + B(n - 1)^2 + C(n - 1)^3 + n^2$$

Equating the coefficients of n, n^2 and the constant term on either side we get $A = \frac{1}{6}$, $B = \frac{1}{2}$, and $C = \frac{1}{3}$. Thus

$$a_n = \frac{1}{6}n + \frac{1}{2}n^2 + \frac{1}{3}n^3 = \frac{n(n + 1)(2n + 1)}{6}$$

which is of course equal to the sum of the squares of the first n natural numbers.

Example 3.3.5

Solve $a_n = a_{n-1} + 12n^2$, where $a_0 = 5$.

Solution.   We see $a_n = a_0 + (12)(n)(n + 1)(2n + 1)/6$ from Example 3.3.4. Thus, $a_n = 5 + 2n(n + 1)(2n + 1)$.

3.4 RECURRENCE RELATIONS AND GENERATING FUNCTIONS

In many instances the nth term a_n in a recurrence relation can be obtained as the coefficient of x^n in the power series expansion of a function $g(x)$ which may be considered as the generating function for the given recurrence relation. Quite often the functional equation for $g(x)$ can be solved algebraically and then a_n is obtained by expressing $g(x)$ as a power series. In other words, the recurrence relation is solved by means of an associated generating function.

Example 3.4.1

Solve the recurrence relation $a_n = 2a_{n-1}$ by using the associated generating function.

Solution.   Let $g(x) = a_0 + a_1x + \cdots + a_nx^n + \cdots$ be the associated generating function. Multiplying both sides of the recurrence relation by x^n, we have $a_nx^n = 2a_{n-1}x^n$, where $n = 0, 1, 2, \ldots$. Thus

$$a_1x = 2a_0x$$

$$a_2x^2 = 2a_1x^2$$

$$\vdots$$

$$a_nx^n = 2a_{n-1}x^n$$

On adding we see that $g(x) - a_0 = 2xg(x)$, which is a functional equation for $g(x)$, and this equation can be solved for $g(x)$. We get $g(x) = a_0/(1 - 2x)$. Thus $a_n = a_02^n$, which is the coefficient of x^n in the power series expansion of $g(x)$.

Example 3.4.2

Solve $a_n = 2a_{n-1} - (n/3)$, where $a_0 = 1$.

Solution.   Since $a_0 = 1$, the associated generating function is

$$g(x) = 1 + a_1x + a_2x^2 + \cdots$$

By putting $n = 1, 2, 3, \ldots$ in $a_nx^n = 2a_{n-1}x^n - (n/3)x^n$, we have

$$a_1 x = 2x - \tfrac{1}{3}x$$

$$a_2 x^2 = 2a_1 x^2 - \tfrac{2}{3}x^2$$

$$\vdots$$

Thus on adding these equations, we have

$$g(x) - 1 = 2xg(x) - \frac{x}{3}(1 + 2x + 3x^2 + \cdots)$$

$$(1 - 2x)g(x) = 1 - \frac{x}{3}f(x)$$

where

$$f(x) = 1 + (1 + 1)x + (1 + 1 + 1)x^2 + \cdots = \frac{1}{(1 - x)^2}$$

[Recall that if $u(x)$ is the ordinary generating function for p_r, then $u(x)/(1 - x)$ is the ordinary generating function for $q_r = p_1 + p_2 + \cdots + p_r$. In the present case $1/(1 - x)$ is the generating function for $p_r = 1$. So $1/(1 - x)^2$ is the generating function for $q_r = 1 + 1 + \cdots + 1 = r$.] Thus from the functional equation for $g(x)$ we get

$$g(x) = \frac{3 - 7x + 3x^2}{3(1 - x)^2(1 - 2x)}$$

Now the function on the right-hand side can be expanded by the method of partial fractions. We then have

$$g(x) = \left(\frac{1}{3}\right)\left[\frac{1}{1 - x} + \frac{1}{(1 - x)^2} + \frac{1}{1 - 2x}\right]$$

Thus

$$a_n = \text{coefficient of } x^n \text{ in } g(x)$$

$$= \frac{1}{3}[1 + (n + 1) + 2^n] = \frac{n + 2 + 2^n}{3}$$

Example 3.4.3

Find the generating function for the recurrence relation $a_n = c_1 a_{n-1} + c_2 a_{n-2}$ with $a_0 = A_0$ and $a_1 = A_1$.

Solution.   The generating function is

$$g(x) = a_0 + a_1 x + a_2 x^2 + \cdots$$

Putting $n = 2, 3, 4, \ldots$ in $a_n x^n = c_1 a_{n-1} x^n + c_2 a_{n-2} x^n$, we have

$$a_2 x^2 = c_1 a_1 x^2 + c_2 a_0 x^2$$

$$a_3 x^3 = c_1 a_2 x^3 + c_2 a_1 x^3$$

$$\vdots$$

On addition, we have

$$g(x) - A_0 - A_1 x = (c_1 x)[g(x) - A_0] + (c_2 x^2) g(x)$$

So $g(x) = u(x)/v(x)$, where $u(x) = (A_0 + A_1 x) - (A_0)(c_1 x)$ and $v(x) = 1 - c_1 x - c_2 x^2$.

Notice the relation between the coefficients of the denominator $v(x)$ and the coefficients of the characteristic function $p(x) = x^2 - c_1 x - c_2$. We can generalize this observation for a linear homogeneous recurrence relation of order r with constant coefficients as follows.

THEOREM 3.4.1

If $p(x)$ and $g(x)$ are the characteristic function and the generating function of the linear homogeneous recurrence relation with constant coefficients $a_n = c_1 a_{n-1} + c_2 a_{n-2} + \cdots + c_r a_{n-r}$ with consecutive initial conditions $a_i = A_i$ ($i = 0, 1, 2, \ldots, r - 1$), then $p(x) = x^r - c_1 x^{r-1} - c_2 x^{r-2} - \cdots - c_r$ and $g(x) = u(x)/v(x)$, in which the denominator is $1 - c_1 x - c_2 x^2 - \cdots - c_r x^r$ and the numerator is

$$[A_0 + A_1 x + \cdots + A_{r-1} x^{r-1}] - (c_1 x)[A_0 + A_1 x + \cdots + A_{r-2} x^{r-2}]$$
$$- (c_2 x^2)[A_0 + A_1 x + \cdots + A_{r-3} x^{r-3}] - \cdots - (c_{r-1} x^{r-1})[A_0]$$

Example 3.4.4

If $p(x) = x^3 - 9x^2 + 26x - 24$, find $g(x)$.

Solution.   The generating function $g(x)$ is $u(x)/v(x)$, where

$$v(x) = 1 - 9x + 26x^2 - 24x^3$$

$$u(x) = (A_0 + A_1 x + A_2 x^2) - (9x)(A_0 + A_1 x) + (26x^2)(A_0)$$

3.5 ANALYSIS OF ALGORITHMS

We first make a distinction between "a problem" and "an instance of the problem." For example, finding the product of two integers is a problem, whereas finding the product of two given integers is an instance of the problem. If the problem is finding the product of two square matrices, then finding the product of two $n \times n$ matrices is an instance of the problem. In network optimization, the shortest-distance problem is the problem of finding the shortest distance from a vertex to the other vertices in a network. An instance of this problem will then be to find the shortest distance from a vertex to the remaining vertices in a given network with n vertices and m edges. In an informal and an intuitive sense, an algorithm for the problem, as we all know, is a step-by-step procedure involving a finite sequence of instructions that can be used to solve every instance of the problem.

From an informal point of view, the **computational complexity of an algorithm for a problem** is the cost, measured in running time or storage or whatever units are relevant, of using the algorithm to solve the problem. This cost, denoted by $f(n)$, depends on the input size n of an instance of the problem. The function f is the complexity function of the algorithm. In the analysis of algorithms, this cost requirement is usually expressed in terms of the number of elementary computational steps, such as arithmetic operations, comparisons, and so on, needed for execution of the algorithm on a (hypothetical) computer on the assumption that all these kinds of operations require unit time. It is quite possible that the cost requirements for two different instances with the same input size could differ significantly. So we consider all instances of a given input size n and then choose the cost requirement of that instance for which the cost is a maximum. In other words, we are examining the worst-case behavior of the algorithm. If A is an algorithm to solve a problem, we denote by $f(A, n)$ the **worst-case complexity** of the algorithm, where n is the input size. For example, consider the problem of choosing the smallest number from a set of n numbers. If we have no information about these numbers and if we choose the smallest number after making all possible comparisons, the number of comparisons is $(n - 1)$ and therefore the worst-case complexity for this procedure of choosing is $(n - 1)$.

As another example consider the ordinary matrix multiplication algorithm, with which we are all familiar. Let L and M be any two $n \times n$ matrices such that no element in either matrix is zero. We thus have a worst-case scenario here. In finding the product matrix we shall not worry about additions. Instead, we count the total number of multiplications involved in implementing the algorithm. Notice that each element in the product matrix is the product of a row r of L and a column c of M. Both r and c have n nonzero numbers. Thus the product involves n multiplications. Since there are n^2 elements in the product matrix the total number of multiplications is $f(A, n) = n^3$, which is the worst-case complexity for matrix multiplication, where A is the usual matrix multiplication algorithm.

Once an algorithm is specified for a problem, we shall write $f(n)$ instead of $f(A, n)$ to denote the worst-case complexity of the algorithm. On the other hand, if we are comparing two algorithms A and A' for the same problem, we write $f(A, n)$ and $f(A', n)$ to denote the respective complexities and in that case we say that A is more **efficient** than A' if $f(A, n)$ does not exceed $f(A', n)$ for all n when $n \geq n_0$ for some fixed positive integer n_0. If $f(A, n) = 5n^2$ and $f(A', n) = n^3$, then A is more efficient than A' since $f(A, n) \leq f(A', n)$ for all $n \geq 5$.

In many cases it is possible to divide a problem into several smaller nonoverlapping subproblems of approximately equal size, solve these subproblems, and merge their solutions to obtain the solution of the original problem. This strategy of solving a problem, known as the **divide-and-conquer** technique, is often more efficient than the usual straightforward method. If $f(A, n)$ is the computational complexity of such a divide-and-conquer algorithm (where n is the input size of an instance of the problem) we have a recurrence relation expressing $f(A, n)$ in terms of $f(A, m)$, where m is the input size of an instance of the subproblem. On solving this recurrence relation using the initial conditions we obtain the complexity function of the algorithm. We now turn our attention to an analysis of some of these recurrence relations that result from such divide-and-conquer algorithms.

Example 3.5.1

If a and b are two n-digit numbers to obtain the product ab, we must perform at most n^2 single-digit multiplications. In other words, if we use the usual multiplication algorithm, its computational complexity is n^2 if we ignore the number of additions and take into account only the total number of multiplications. We can use a more efficient divide-and-conquer algorithm to multiply the two numbers as follows.

Assume that n is even. Then both a and b can be subdivided into two parts:

$$a = a_1(10)^{n/2} + a_2$$

$$b = b_1(10)^{n/2} + b_2$$

where the parts a_1, a_2, b_1, and b_2 are all $(n/2)$-digit numbers. Then

$$ab = a_1 b_1(10)^n + [a_1 b_2 + a_2 b_1](10)^{n/2} + a_2 b_2$$

The expression on the right involves four $(n/2)$-digit multiplications. We can reduce this into three multiplications if we make use of the formula

$$(a_1b_2 + a_2b_1) = (a_1 + a_2)(b_1 + b_2) - a_1b_1 - a_2b_2$$

Thus we have the recurrence relation $f(n) = 3f(n/2)$ with $f(1) = 1$. Of course, there is a possibility that $(a_1 + a_2)$ or $(b_1 + b_2)$ may be $(n/2 + 1)$-digit numbers, but this does not affect complexity considerations. If we write $n = 2^k$ in the recurrence relation, we get

$$f(n) = 3f\left(\frac{n}{2}\right) = 3^2f\left(\frac{n}{2^2}\right) = \cdots = 3^kf(1) = 3^{\log n} = n^{\log 3}$$

Thus the complexity of this recursive algorithm is $n^{\log 3}$, which is less than n^2 for all $n \geq 1$.

More generally, let us consider a divide-and-conquer algorithm that splits a problem of size n into several subproblems each of size n/b. We shall assume that $n = b^k$ and that the number of subproblems is a, where $a > 1$. So if the complexity of the algorithm for the problem is $f(n)$, then the complexity of the same algorithm for the subproblem is $f(n/b)$. We thus have a recurrence relation of the form $f(n) = af(n/b) + h(n)$, $f(1) = d$, $a > 1$, $n = b^k$, where $h(n)$ represents the cost of dividing the problem into subproblems plus the cost of merging the solutions of these subproblems to obtain the solution of the original problem. Two special cases are of particular importance: (1) $h(n)$ is a constant c and (2) $h(n) = cn$, where c is a constant.

Example 3.5.2

Solve $f(n) = af(n/b) + c$, $f(1) = d$, where a and b are integers that are at least 2, $n = b^k$, and c is a constant.

Solution

$$f(n) = af\left(\frac{n}{b}\right) + c = a\left[af\left(\frac{n}{b^2}\right) + c\right] + c$$

$$= a^2f\left(\frac{n}{b^2}\right) + c[1 + a]$$

$$\vdots$$

$$= a^kf\left(\frac{n}{b^k}\right) + c[1 + a + a^2 + \cdots + a^{k-1}]$$

$$= a^kd + c\frac{(a^k - 1)}{a - 1}$$

$$= Aa^k + B$$

where $A = (ad - d + c)/(a - 1)$ and $B = (-c)/(a - 1)$. Since $n = b^k$, we have $k = \log_b n$ and $a^k = n^r$, where $r = \log_b a$. Thus $f(n) = An^r + B$.

Example 3.5.3

Solve $f(n) = af(n/b) + cn$ with the same conditions as in Example 3.5.2.

Solution.   By iteration we have

$$f(n) = a^k f(1) + cn\left[1 + \frac{a}{b} + \left(\frac{a}{b}\right)^2 + \cdots + \left(\frac{a}{b}\right)^{k-1}\right]$$

Case (1): $a = b$

$$f(n) = a^k d + cnk = b^k d + cnk = nd + cn\log_b n$$

Thus $f(n) = c(n \log n) + d(n)$.

Case (2): $a < b$ or $a > b$. After some simplification we get $f(n) = An^r + Bn$, where $A = (bd - ad - bc)/(b - a)$ and $B = bc/(b - a)$ and $r = \log_b a$. If $a < b$, then $r < 1$, so Bn is the dominating term in $f(n)$. If $a > b$, then $r > 1$, so An^r is the dominating term in $f(n)$.

Matrix Multiplication

We next consider the efficiency of matrix multiplication algorithms. As we noticed earlier, if A and B are two 2×2 matrices, we usually perform at least eight ($2^3 = 8$) multiplications to obtain the product matrix $C = AB$. By a clever algebraic manipulation, it is possible to reduce the number of multiplications from eight to seven. Let $a(i, j)$ denote the element in A at the ith row and jth column; similarly for B and C.

Now define the following seven products:

$$x_1 = [a(1, 1) + a(2, 2)] \cdot [b(1, 1) + b(2, 2)]$$

$$x_2 = [a(2, 1) + a(2, 2)] \cdot b(1, 1)$$

$$x_3 = a(1, 1) \cdot [b(1, 2) - b(2, 2)]$$

$$x_4 = a(2, 2) \cdot [b(2, 1) - a(2, 1)]$$

$$x_5 = [a(1, 1) + a(1, 2)] \cdot b(2, 2)$$

$$x_6 = [a(2, 1) - a(1, 1)] \cdot [b(1, 1) + b(1, 2)]$$

$$x_7 = [a(1, 2) - a(2, 2)] \cdot [b(2, 1) + b(2, 2)]$$

Then it is easy to verify that every element $c(i, j)$ of the product matrix can be expressed as a sum or difference of some of these seven numbers as follows:

$$c(1, 1) = x_1 + x_4 - x_5 + x_7 \qquad c(1, 2) = x_3 + x_5$$

$$c(2, 1) = x_2 + x_4 \qquad\qquad\qquad c(2, 2) = x_1 + x_3 - x_2 + x_6$$

Thus we obtain the product matrix by performing at most seven multiplications instead of eight. We utilize this information to define a divide-and-conquer algorithm for matrix multiplication and solve the associated recurrence relation to obtain the computational complexity of this algorithm in the next example.

Example 3.5.4

Let A and B be two $n \times n$ matrices where $n = 2^k$. Obtain a divide-and-conquer algorithm to obtain the product matrix AB and find the complexity of the this algorithm.

Solution.   We partition each matrix into four submatrices where each submatrix is a $(n/2) \times (n/2)$ matrix. See Figure 3.5.1.

The four submatrices of A are: (1) $A(1, 1)$ obtained by considering the first $n/2$ rows and the first $n/2$ columns of A; (2) $A(1, 2)$ obtained by considering the first $n/2$ rows and the last $n/2$ columns; (3) $A(2, 1)$ obtained by considering the last $n/2$ rows and the first $n/2$ columns; and (4) $A(2, 2)$ obtained by considering the last $n/2$ rows and the last $n/2$ columns of A; similarly for B and C.

We now define the following seven submatrices:

$$X_1 = [A(1, 1) + A(2, 2)] \cdot [B(1, 1) + B(2, 2)]$$

$$X_2 = [A(2, 1) + A(2, 2)] \cdot B(1, 1)$$

$$X_3 = A(1, 1) \cdot [B(1, 2) - B(2, 2)]$$

$$X_4 = A(2, 2) \cdot [B(2, 1) - B(1, 1)]$$

$$X_5 = [A(1, 1) + A(1, 2)] \cdot B(2, 2)$$

$$X_6 = [A(2, 1) - A(1, 1)] \cdot [B(1, 1) + B(1, 2)]$$

$$X_7 = [A(1, 2) - A(2, 2)] \cdot [B(2, 1) + B(2, 2)]$$

Notice that each of these $(n/2) \times (n/2)$ submatrices is a product of two $(n/2) \times (n/2)$ submatrices. It can easily be verified that

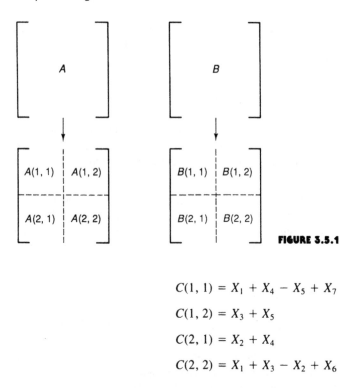

FIGURE 3.5.1

$$C(1, 1) = X_1 + X_4 - X_5 + X_7$$

$$C(1, 2) = X_3 + X_5$$

$$C(2, 1) = X_2 + X_4$$

$$C(2, 2) = X_1 + X_3 - X_2 + X_6$$

Thus if $f(n)$ is the number of multiplications needed to multiply A and B using this divide-and-conquer strategy we have the recurrence relation $f(n) = 7f(n/2)$ with the initial condition $f(1) = 1$. On solving this we get $f(n) = 7^k f(1) = 7^k$, where $k = \log_2 n$. Equivalently, $f(n) = n^r$, where $r = \log_2 7$. Thus the complexity of this algorithm is less than n^3, so this method is more efficient than the usual matrix multiplication method!

Evaluation of a Polynomial at a Given Point

We conclude our algorithm analysis with a discussion on the number of multiplications and additions involved to evaluate the value of a polynomial $p(x)$ for a given value of x. If the degree of $p(x)$ is n, it is easy to see that a straightforward nonrecursive evaluation of $p(x)$ at $x = t$ will involve $(2n - 1)$ multiplications and n additions. For example, to evaluate $p(x) = 5x^3 + 8x^2 + 4x + 7$ at $x = 27$, we perform the five multiplications (1) $a = 27 \cdot 27$, (2) $b = a \cdot 27$, (3) $c = 4 \cdot 27$, (4) $8 \cdot a$, and (5) $5 \cdot b$ and then three additions. A more efficient (recursive) method known as the Horner's method (or Newton's method) requires only n multiplications and n additions. If we write the polynomial in a telescoping form as $p(x) = (((5 \cdot x + 8) \cdot x) + 4) \cdot x + 7$, the number of multiplications is only 3.

Example 3.5.5 (Horner's Method)

(a) If $f(n)$ is the number of multiplications needed to evaluate a polynomial of degree n at a point, obtain a recursive relation for $f(n)$.

(b) If $g(n)$ is the total number of multiplications and additions, find a recursive relation involving $g(n)$.

Solution. If $p(x)$ is a polynomial of degree n, then $p(x) = xq(x) + a$, where a is a constant of $q(x)$ is a polynomial of degree $(n - 1)$. (a) An evaluation of $q(x)$ at a point will involve $f(n - 1)$ multiplications. Thus $f(n) = f(n - 1) + 1$, with the initial condition $f(0) = 0$. Thus $f(n) = n$.

(b) Obviously, $g(n) = g(n - 1) + 2$ with $g(0) = 0$. The solution is $g(n) = 2n$.

Finally, we discuss another recursive algorithm for polynomial evaluation, which is more efficient than Horner's method. [This method is explained in detail in Baase (1978).] First a definition and some properties of polynomials. A polynomial $p(x)$ is called a **monic polynomial** if the coefficient of the leading term is 1. We also assume that the degree n of $p(x)$ is equal to $2^k - 1$. The number of multiplications needed for polynomial evaluation for $p(x)$ is $(n - 1) = 2^k - 2$ by Horner's method.

Can we do better regarding the number of multiplications? Now it can be shown that if $p(x) = x^n + a_{n-1}x^{n-1} + \cdots + a_1x + a_0$ is a monic polynomial of degree $2^k - 1$, we can write $p(x)$ as $p(x) = (x^j + b) \cdot q(x) + r(x)$ where (1) both $q(x)$ and $r(x)$ are both monic polynomials of degree $(n - 1)/2$ and $j = (n + 1)/2$, (2) $b = a_{j-1} - 1$, and (3) the coefficients of $q(x)$ are the first $(n + 1)/2$ coefficients of $p(x)$, starting from the highest term. If a power of x is missing, the corresponding coefficient is taken as zero. For example, let

$$p(x) = x^7 + 2x^6 + 2x^5 + 3x^4 + 9x^3 + 9x^2 + 18x + 9$$

Here $2^k - 1 = 7$. So $k = 3$ and $j = 4$. Also, $b = a_3 - 1 = 9 - 1 = 8$. Thus $p(x) = (x^4 + 8)q(x) + r(x)$, where $q(x)$ is a polynomial of degree 3 whose coefficients are 1, 2, 2, 3 starting from the highest term. In other words, $q(x) = x^3 + 2x^2 + 2x + 3$ and

$$r(x) = p(x) - (x^4 + 8)q(x) = x^3 - 7x^2 + 2x - 15$$

Proceeding similarly we write

$$q(x) = (x^2 + 1)(x + 2) + (x + 1)$$

and

$$r(x) = (x^2 + 1)(x - 7) + (x - 8)$$

Thus finally, we have

$$p(x) = (x^4 + 8) \cdot [(x^2 + 1) \cdot (x + 2) + (x + 1)]$$
$$+ (x^2 + 1) \cdot (x - 7) + (x - 8)$$

which involves five multiplications in all (three as shown with parentheses and two for computing x^2 and x^4). On the other hand, if we use Horner's method, the number of multiplications will be six instead of five. These observations are generalized as follows.

THEOREM 3.5.1

Let $p(x)$ be a monic polynomial of degree n, where $n = 2^k - 1$. Then the number of multiplications needed to evaluate $p(x)$ at a point is at most $(n - 3)/2 + \log(n + 1)$, and the number of additions/substractions needed is at most $(3n - 1)/2$.

Proof:

Let $n = 2^k - 1$. Also let $f(k)$ be the number of multiplications needed to evaluate $p(x)$ (without taking into account the number of multiplications required to compute the various powers of x at the given point), and let $g(k)$ be the number of additions needed. Since $p(x) = (x^j + b)q(x) + r(x)$, we have the recurrence relation $f(k) = 2f(k - 1) + 1$ with the initial condition $f(1) = 0$ and the relation $g(k) = 2g(k - 1) + 2$ with $g(1) = 1$.

On solving these recurrence relations, we have

$$f(k) = 2^{k-1} - 1 = \frac{n - 1}{2} \quad \text{and} \quad g(k) = 3 \cdot 2^{k-1} - 2 = \frac{3n - 1}{2}$$

Next we consider the number of multiplications involved to compute the various powers of x. We have to compute x^2, then x^4, then x^8, and so on, until we reach the jth power of x, where $j = (n + 1)/2 = 2^{k-1}$, and this process will involve $(k - 1)$ multiplications. But $k = \log(n + 1)$. Thus this algorithm will need at most $(n - 1)/2 + \log(n + 1) - 1 = (n/2) + \log(n + 1) - (3/2)$ multiplications and $(3n - 1)/2$ additions, whereas Horner's method will need at most $(n - 1)$ multiplications and n additions. $\diamondsuit$

3.6 NOTES AND REFERENCES

The pioneering work in the study of recurrence relations was done by Leonardo of Pisa (more popularly known as Fibonacci) in the thirteenth century and subsequently by Jacob Bernoulli (1654–1705), his nephew Daniel Bernoulli (1692–1770), James Stirling (1692–1770), and Leonhard Euler (1707–1783). Some useful references are the relevant chapters in the books on combinatorics and discrete mathematics by Cohen (1978), Grimaldi (1985), Krishnamurthy (1986), Liu (1968), Liu (1985), Roberts (1984), Townsend (1987), and Tucker (1984). The techniques for solving recurrence relations were first used in the development of the theory of difference equations. See Levy and Lessman (1961) for a survey of these methods. An excellent reference for the applications of difference equations is Goldberg (1958). For additional reading on the analysis of algorithms, see the relevant chapters in the books by Baase (1978), Knuth (1973), Roberts (1984), Stanat and McAllister (1977), and Wilf (1986).

3.7 EXERCISES

3.1. Suppose that there are n lines in a plane such that no two lines are parallel and no three lines are concurrent dividing the plane into $f(n)$ distinct regions. Find a recurrence relation for $f(n)$ and solve for $f(n)$. Find the value of $f(9)$.

3.2. Find a recurrence relation for $f(n)$, the number of n-letter words that can be formed using the letters A, B, C, D, and E such that in each word the frequency of the letter A is an odd number.

3.3. (The Tower of Hanoi Problem.) Three vertical cylindrical poles of equal radius and height are placed along a line on top of a table and n circular disks of decreasing radius, each with a hole at its center, are attached to the first pole such that the largest one is at the bottom, the next largest is just above that, and so on, and finally the smallest is on top the heap (Figure 3.7.1). The distance between the feet of any two poles is not less than the diameter of the largest disk. A legal move is defined as the transfer of the top disk from any one of the three poles to another pole such that no disk is placed on top of a smaller disk. Let $f(n)$ be the number of legal moves required to transfer all the disks from the first pole to one of the other two poles. Obtain a recurrence relation for $f(n)$ and solve this relation.

3.4. In climbing up a staircase, an ordinary step covers at least one stair and at most two stairs. If $f(n)$ is the number of ways of climbing up a staircase (making only ordinary steps) with n stairs, find a recurrence relation for $f(n)$.

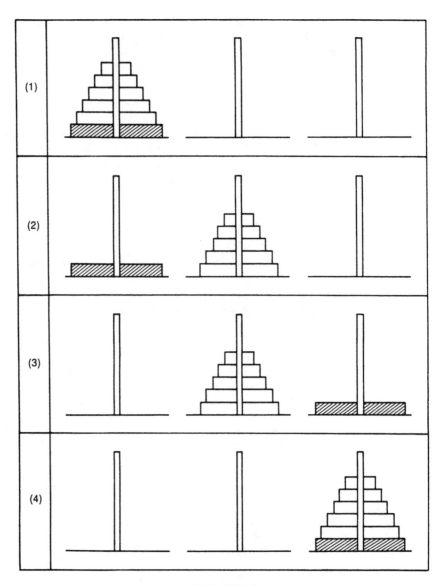

FIGURE 3.7.1

3.5. Let S be the set of all binary words of length n such that two zeros do not appear consecutively in any word in the set. Find a recurrence relation for the number of elements in S.

3.6. There are n personal checks made out in the name of n individuals whose names appear on n envelopes. Find a recurrence relation for the number of ways these checks can be placed in the n envelopes so that no check is in the right envelope.

3.7. Show that the recurrence relation $f(n) = (n - 1)[f(n - 1) + f(n - 2)]$ with $f(1) = 0$ and $f(2) = 1$ can be simplified as $f(n) = nf(n - 1) + (-1)^n$ with $f(2) = 1$.

3.8. Let $f(n)$ be the number of subsets of a set that has n elements. Find a recurrence relation for $f(n)$.

3.9. Let $f(n)$ be the number of elements in X that is the set of all n-symbol words formed by the symbols A, B, and C such that no word has a pair of consecutive A's.

3.10. Solve for k in the recurrence relation $f(n + 1) = kf(n)$ if **(a)** $f(1) = 5$ and $f(2) = 10$, and **(b)** $f(1) = 5$ and $f(3) = 20$.

3.11. Solve: $f(n + 3) = 6f(n + 2) - 11f(n + 1) + 6f(n)$, where $f(0) = 3$, $f(1) = 6$, and $f(2) = 14$.

3.12. Solve: $f(n + 3) = 4f(n + 2) - 5f(n + 1) + 2f(n)$, where $f(0) = 2$, $f(1) = 4$, and $f(2) = 7$.

3.13. Solve: $f(n + 3) = 3f(n + 2) + 4f(n + 1) - 12f(n)$ where $f(0) = 0$, $f(1) = -11$, and $f(2) = -15$.

3.14. The roots of the characteristic equation of a linear homogeneous recurrence relation with constant coefficients are 1, 2, 2, and 3. Write down the relation and its general solution.

3.15. Solve: $nf(n) - (5n - 5) f(n - 1) = 0$ where $f(1) = 10$. [*Hint*: Substitute $g(n) = nf(n)$.]

3.16. Let A be the $m \times m$ matrix in which all the diagonal numbers are equal to 0 and all the nondiagonal numbers are equal to 1. Then the diagonal numbers of A^n are all equal to a positive integer $f(n)$ and the nondiagonal numbers of A^n are all equal to a positive integer $g(n)$ for any positive integer n. Prove that $f(n + 1) = (m - 1)g(n)$ and $g(n + 1) = f(n) + (m - 2)g(n)$ and use this fact to obtain a recurrence relations for $g(n)$ with appropriate initial conditions. Solve the relation. Find $g(n)$ and $f(n)$.

3.17. Solve the following inhomogeneous recurrence relations involving $f(n)$: $f(n) - 4f(n - 1) + 4f(n - 2) = h(n)$ where **(a)** $h(n) = 1$, **(b)** $h(n) = n$, **(c)** $h(n) = 3^n$, **(d)** $h(n) = 2^n$, and **(e)** $h(n) = 1 + n + 2^n + 3^n$.

3.18. Solve the relation $f(n + 2) - 4f(n + 1) + 3f(n) = 16$ with the initial condition $f(0) = 4$ and $f(1) = 2$.

3.19. Solve: $f(n) = 4f(n - 1) + 5(3)^n$.

3.20. Solve: $f(n) = 4f(n - 1) + 5(4)^n$.

3.21. Solve: $f(n) = f(n - 1) + 2f(n - 2) + 4(3)^n$ with the initial conditions $f(0) = 11$ and $f(1) = 28$.

3.22. Solve: $f(n) = 4f(n - 1) - 4f(n - 2) + (2)^n$.

3.23. Solve the recurrence relation $f(n) = f(n - 1) + 6n^2$, $f(0) = 0$, by **(a)** using the characteristic root, and **(b)** repeated substitution. Hence find the sum of the squares of the first n natural numbers.

3.24. Find the constants p, q, and r in the recurrence relation $f(n) + pf(n - 1) + qf(n - 2) = r$ if $f(n) = A(2)^n + B(3)^n + 4$.

3.25. If the ordinary generating function of a recurrence relation involving $f(n)$ is $g(x) = (2)/[(1 - x)(1 - 2x)]$, find $f(n)$.

3.26. Solve the recurrence relation $f(n) = f(n - 2) + 4n$, $f(1) = 2$, $f(0) = 3$, by using the appropriate ordinary generating function.

3.27. Prove that if $g(x)$ is the ordinary generating function for the recurrence relation $f(n + 1) = (n + 1)f(n) + (-1)^{n+1}$ with the initial condition $f(0) = 1$, then $g(x)$ satisfies the differential equation $g'(x) + [(x - 1)/x^2]g(x) + (1)/(x^2)(1 + x) = 0$. Solve the recurrence relation by using its exponential generating function.

3.28. The recurrence relation for a divide-and-conquer algorithm is $f(n) = 9f(n/3) + 8n$ with the initial condition $f(1) = 1$, where $n = 3^r$. Solve for $f(n)$ as a function of n.

3.29. Solve the recurrence relation $f(n) = 5f(n/2) - 6f(n/4) + n$ with the initial conditions $f(1) = 2$ and $f(2) = 1$, where $n = 2^r$.

3.30. Solve $f(n) = f(n/b) + c$ with the initial condition $f(1) = d$, where $n = b^r$.

3.31. Find the ordinary generating function for the recurrence relation $f(n + 1) = af(n) + b^n$ with the initial condition $f(0) = c$, where a, b, and c are constants.

3.32. Consider the example in this section that discusses the complexity of matrix multiplications. If $f(n)$ is the total number of multiplications involved in finding the product of two $n \times n$ matrices, it was proved that $f(n) = n^r$, where $r = \log 7$. Find a recursive relation for the number of additions involved and solve it.

3.33. Find the recurrence relation (involving the number of comparisons) for the divide-and-conquer algorithm to find the largest and smallest elements in a set of n numbers and solve it.

3.34. Let $f(n)$ be the number of comparisons required to sort a list of n numbers in nondecreasing order. **(a)** Obtain a recursive relation expressing $f(n)$ in terms of $f(n - 1)$ with appropriate initial condition. **(b)** Obtain a recursive relation that expresses $f(n)$ in terms of $f(n/2)$ with appropriate initial condition. **(c)** Solve these two recurrence relations and compare the efficiency of the two algorithms involved.

CHAPTER

4

Graphs and Digraphs

4.1 INTRODUCTION

Even though the origins of graph theory can be traced back to the days of the great Swiss mathematician Leonhard Euler (1707–1783), only since the 1930s has there been a sustained and intense interest in graph theory as a mathematical discipline. These days, graph theory is one of the most popular and fertile branches in mathematics and computer science. One important reason for this revived and renewed interest in graph theory is its applicability to many of the complex and wide-ranging problems of modern society in such diverse fields as economics, facility location, management science, marketing, energy modeling, transmission of information, and transportation planning to name a few. Quite often such problems can be modeled as graphs or networks. In this respect graph theory is used first and foremost as a tool for formulating problems and defining structural interrelationships. Once a problem is formulated in graph-theoretical language, it becomes relatively easy to comprehend it in its generality. The next step will, of course, be to explore avenues to seek a solution to the problem. The field of graph theory has two different branches: the algebraic aspects and the optimization aspects. In Chapters 4, 5, and 6 we discuss the former. The area of network optimization, which is greatly advanced by the advent of the computer, is the topic for the last two chapters.

A **graph** $G = (V, E)$ is a structure consisting of a finite set V of **vertices** (also known as the **nodes**) and a finite set E of **edges** such that each edge e is associated with a pair of vertices v and w. We write $e = \{v, w\}$ or $\{w, v\}$ and say that (1) e is an edge **between** v and w, (2) e is **incident** on both v and w, and (3) e **joins** the vertices v and w. In this case both v and w are **adjacent vertices** and they are **incident** on e. An edge joining a vertex to itself is a **loop**. If there is more than one edge joining pairs of vertices in a graph, the graph is a **multigraph.** If two or more edges join the same pair of vertices in a multigraph, these edges are called **multiple edges.** In a pictorial representation of a graph, a vertex is drawn as a small circle with the name (or number) of the vertex written inside the circle. An edge between two vertices is represented by a segment of a line or a curve joining the two circles that represent the vertices. In Figure 4.1.1 we have a pictorial representation of a multigraph in which the vertex set is $V = \{1, 2, 3, 4, 5, 6, 7, 8, 9\}$ with loops at vertex 2 and at vertex 8. There are two edges between 7 and 9 and three edges between 4 and 5. A graph is **simple** if it has no loops or multiple edges. If a real number is associated with each edge, then G is a **network** or a **weighted graph.**

A **directed graph** or **digraph** is a structure $G = (V, E)$ where again V is a finite set of vertices and E is finite set of **arcs** such that each arc e in E is associated with an *ordered pair* of vertices v and w. We write $e = (v, w)$ and we say that (1) e is an arc from v to w, (2) vertex v is **adjacent** to vertex w, (3) vertex w is **adjacent from** vertex v, (4) arc e is **incident from** v, and (5) arc e is **incident to** w. Two vertices are **adjacent** if there is an arc from one to the other. We have a **weighted digraph** or **directed network** whenever a real number

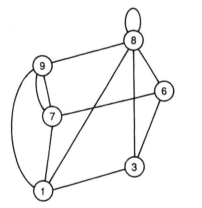
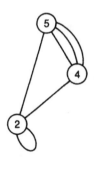

FIGURE 4.1.1

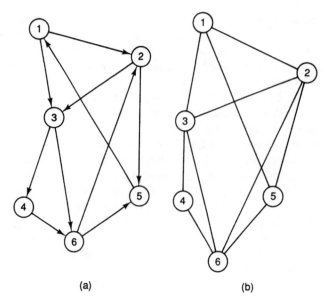

(a)   (b)   **FIGURE 4.1.2**

is associated with each arc. If we treat every arc of a digraph as an edge, the resulting structure is called the **underlying graph** of the digraph. In a pictorial representation of a digraph an arc from vertex v to vertex w is drawn as a directed segment with the arrowhead pointing toward w. In Figure 4.1.2(a) we have a pictorial representation of a digraph the underlying graph of which is as in Figure 4.1.2(b). In a **mixed graph** $G = (V, E)$ at least one element of E is an arc and at least one element is an edge. An element in the former category is a directed arc, whereas an element in the latter is an undirected edge. The real numbers associated with the edges and arcs in networks are usually written along these edges and arcs.

Of special interest is the **bipartite graph** the vertices of which can be partitioned into two disjoint sets V and W such that each edge is an edge between a vertex in V and a vertex in W and is denoted by $G = (V, W; E)$. In Figure 4.1.3 we have a pictorial representation of a bipartite graph in which $V =$

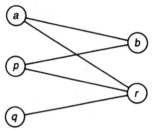

FIGURE 4.1.3

$\{a, p, q\}$ and $W = \{b, r\}$. $G = (V, W; E)$ is a **bipartite digraph** if every arc in E is from a vertex in V to a vertex in W.

A simple graph with n vertices is **complete** if there is an edge between every pair of vertices. The graph is then denoted by K_n. A digraph is a **complete digraph** if its underlying graph is complete. A simple bipartite graph $G = (V, W, E)$ is a **complete bipartite graph** if there is an edge between every vertex in V and every vertex in W. The bipartite graph then is denoted by $K_{p,q}$ if there are p vertices in V and q vertices in W.

A graph $G' = (V', E')$ is a **subgraph** of $G = (V, E)$ if V' is a subset of V and E' is a subset of E. If W is any subset of V, the **subgraph of G induced by W** is the graph $H = (W, F)$ where f is an edge in F if $f = \{u, v\}$, where f is in E and both u and v are in W. In Figure 4.1.4, $W = \{1, 2, 4, 5\}$ is a subset of the vertex set V of the graph G and the subgraph of G induced by W is H. A complete subgraph of G is called a **clique** in G.

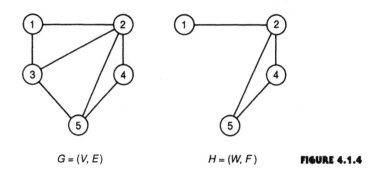

$G = (V, E)$ $H = (W, F)$ **FIGURE 4.1.4**

Example 4.1.1   (The Königsberg Bridge Problem)

The first publication in graph theory is that of Leonhard Euler in 1736. His paper presented a solution to what is known as the Königsberg bridge problem. The city of Königsberg (now known as Kaliningrad) in Russia, situated by the Pregel River, consists of the north shore (N), the south shore (S), the west island (W), and the east island (E). Linking these four parts were seven bridges: two between N and W, two between S and W, and one each from E to N, S, and W. (See Figure 4.1.5.) The problem posed to Euler was whether it is possible to start from any location in the city and return to the starting point after crossing each bridge exactly once. If each part of the city is considered as a vertex and if each bridge is considered as an edge, we have a graph with four vertices and seven edges (see Figure 4.1.6), giving a graph model of the problem that can be stated as follows: Given a graph (not necessarily simple), is it possible to trace the entire diagram of the graph without going over the same edge more

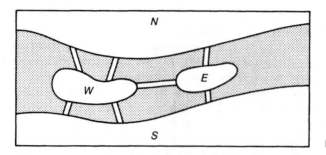

FIGURE 4.1.5

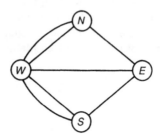

FIGURE 4.1.6

than once? That the answer is no in the case of the Königsberg bridge problem was easily established by Euler. More on this in our discussion of Eulerian graphs in Chapter 5.

Example 4.1.2   (Communication Digraphs)

Consider an organization consisting of several components. Let each component be a vertex. Draw an arrow from vertex v to vertex w if component v can transmit signals to component w. The resulting digraph is known as a communication digraph.

Example 4.1.3   (Transportation Networks)

Suppose that we let each vertex in a graph represent a city in the United States. Two vertices are joined by an edge if there is direct nonstop air service between them. A natural question that arises is whether it is possible to start from a city and return to the starting point after visiting each city exactly once. This problem is discussed in Chapter 5 when we investigate Hamiltonian graphs, named after the nineteenth-century Irish mathematician Sir William Hamilton, who did pioneering work in this area. If a nonnegative real number is assigned to each edge to represent the cost of using that edge, a related optimization problem then is to find such a tour (if it exists) so that the cost of the tour is as small as possible. This is the celebrated traveling salesman problem (TSP), which is a central topic in combinatorial optimization.

Example 4.1.4   (Tournaments)

In a round-robin tennis tournament each player must play every other player and no ties are allowed. Let each vertex in a digraph represent a player. Draw an arrow from vertex v to vertex w if v defeats w. The resulting digraph is complete and it is known as a *dominance graph* of a tournament. Such dominance digraphs arise frequently in social and biological sciences. A basic problem here is to decide who the "winner" or "leader" is in a dominance graph. In Figure 4.1.7 we have a tournament consisting of four players in which the player represented by vertex 2 is the winner. More on tournaments in Chapter 5.

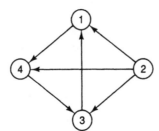

FIGURE 4.1.7

Example 4.1.5   (The Assignment Problem)

Suppose that there are m job applicants $p_1, p_2, \ldots, p_m$ and n jobs $q_1, q_2, \ldots, q_n$. Let V be the set of job applicants and W be the set of jobs. If p_i is qualified for q_j, draw an edge between those two vertices and let $c_{i,j}$ represent the salary to be paid to p_i if she or he is hired for the job q_j. The model we have in this case is a weighted bipartite network, and the optimization problem then is to find a job assignment for the applicants such that (a) all the jobs are filled and (b) the total salary to be paid is a minimum.

4.2 ADJACENCY MATRICES AND INCIDENCE MATRICES

For the purpose of inputting graphs into a computer it is necessary to describe graphs without resorting to their pictorial representations. Moreover, such diagrams are not very practical when graphs with a large number of vertices and

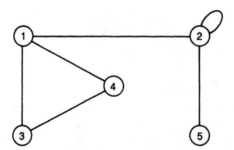

FIGURE 4.2.1

edges are to be studied. There are several ways to represent a graph or a digraph without a pictorial representation, and in this section we discuss some of them. The best way to input a graph depends on its properties and subsequent uses. Furthermore, the efficiency of a graph algorithm depends on the choice of the method for representing the graph under consideration.

Let $G = (V, E)$ be a graph with no multiple edges where $V = \{1, 2, 3, \ldots, n\}$. The **adjacency matrix** of G is the $n \times n$ matrix $A = (a_{ij})$, where $a_{ij} = 1$ if there is an edge between vertex i and vertex j and $a_{ij} = 0$ otherwise. The adjacency matrix of a graph is symmetric. Its diagonal elements are zero if and only if there are no loops. The adjacency matrix of the graph in Figure 4.2.1 is the matrix A, where

$$A = \begin{bmatrix} 0 & 1 & 1 & 1 & 0 \\ 1 & 1 & 0 & 0 & 1 \\ 1 & 0 & 0 & 1 & 0 \\ 1 & 0 & 1 & 0 & 0 \\ 0 & 1 & 0 & 0 & 0 \end{bmatrix}$$

The **degree** of a vertex in a graph is the number of edges incident on that vertex. A vertex is **odd** if its degree is odd; otherwise, it is **even**. In Figure 4.2.1, vertices 1, 2, and 5 are odd, with degrees 3, 3, and 1, respectively. Obviously, the number of nonzero elements in row i of the adjacency matrix of a graph is the degree of vertex i, which is also equal to the sum of all the elements of row i or column i.

The **adjacency matrix of a digraph** with n vertices is also a square $n \times n$ matrix $A = (a_{ij})$, where $a_{ij} = 1$ is there is an arc from i to j and is 0 otherwise. The adjacency matrix of the digraph in Figure 4.2.2 is A, where

$$A = \begin{bmatrix} 0 & 0 & 1 & 1 & 0 \\ 1 & 0 & 0 & 1 & 1 \\ 0 & 1 & 1 & 0 & 0 \\ 0 & 0 & 1 & 0 & 0 \\ 1 & 0 & 1 & 1 & 0 \end{bmatrix}$$

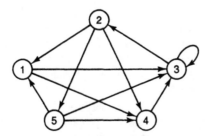

FIGURE 4.2.2

Notice that the adjacency matrix of a digraph need not be symmetric.

The **outdegree** of a vertex in a digraph is the number of arcs incident from that vertex and the **indegree** of a vertex is the number of arcs incident to that vertex. In Figure 4.2.2 the outdegree of vertex 2 is 3 and its indegree is 1. Notice that the sum of the elements in row i is the outdegree of i and the sum of the elements in column j is the indegree of j. Notice also that the outdegree of 3 is 2 and that its indegree is 4.

Another matrix that is useful for entering graphs and digraphs in a computer is the incidence matrix. Unlike the adjacency matrix the incidence matrix is capable of representing multiple edges and parallel arcs. Let $G = (V, E)$ be a graph where $V = \{1, 2, \ldots, n\}$ and $E = \{e_1, e_2, \ldots, e_m\}$. The **incidence matrix** of G is a $n \times m$ graph $B = (b_{ik})$, where each row corresponds to a vertex and each column corresponds to an edge such that if e_k is an edge between i and j, then all elements of column k are 0 except $b_{ik} = b_{jk} = 1$. For example, the incidence matrix of the graph in Figure 4.2.3 is B, where

$$B = \begin{array}{c} \begin{array}{cccccc} e_1 & e_2 & e_3 & e_4 & e_5 & e_6 \end{array} \\ \begin{bmatrix} 1 & 1 & 1 & 1 & 0 & 0 \\ 1 & 0 & 0 & 0 & 1 & 1 \\ 0 & 1 & 1 & 0 & 0 & 0 \\ 0 & 0 & 0 & 0 & 0 & 1 \\ 0 & 0 & 0 & 1 & 0 & 0 \end{bmatrix} \end{array}$$

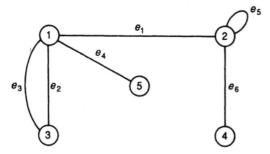

FIGURE 4.2.3

Notice that a column which corresponds to an edge has exactly two nonzero elements if it is not a loop and exactly one nonzero element if it is a loop. Furthermore, the sum of the elements of row i is the degree of vertex i.

We also observe that in any graph with no loops the sum of the degrees of all the vertices is twice the number of edges since each edge is accounted twice, once for each of its incident vertices. For example, in Figure 4.1.6 we see deg N + deg S + deg W + deg E = 3 + 3 + 5 + 3 = 14 = twice the number of edges. We state this property as a theorem that is sometimes known as the *first theorem of graph theory*.

THEOREM 4.2.1

If G is multigraph with no loops and m edges, the sum of the degrees of all the vertices of G is $2m$.

COROLLARY

The number of odd vertices in a loopless multigraph is even.

Proof:

Suppose that the number of odd vertices is r. Let p be the sum of the degrees of all odd vertices and q be the sum of the degrees of all even vertices. Then $p + q$ is even by the theorem. Also, q is even, so p is even. But p is the sum of r odd numbers. Therefore, r is even, which proves the assertion. $\diamondsuit$

The **incidence matrix B of a digraph** (with no loops) is defined as follows: If e_k is an arc from i to j, all elements in column k are zero except $b_{ik} = -1$ and $b_{jk} = 1$. For example, the incidence matrix of the digraph in Figure 4.2.4 is B, where

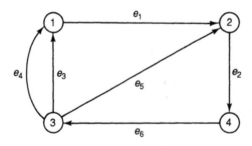

FIGURE 4.2.4

$$B = \begin{array}{c} \\ \\ \\ \\ \\ \end{array} \begin{array}{cccccc} e_1 & e_2 & e_3 & e_4 & e_5 & e_6 \\ \end{array}$$

$$B = \begin{bmatrix} -1 & 0 & 1 & 1 & 0 & 0 \\ 1 & -1 & 0 & 0 & 1 & 0 \\ 0 & 0 & -1 & -1 & -1 & 1 \\ 0 & 1 & 0 & 0 & 0 & -1 \end{bmatrix}$$

Notice that the sum of all the elements in row i of the incidence matrix of a digraph is equal to the indegree of i minus the outdegree of i. We also observe that **in any digraph the sum of all outdegrees is equal to the total number of arcs, which is again equal to the sum of all indegrees.** This is because when the outdegrees are summed, each arc is counted once since every arc is incident from one vertex. Similarly, when the indegrees are summed, each arc is counted once since every arc is incident to a single vertex.

4.3 JOINING IN GRAPHS

A **path between two vertices** v_1 and v_r in a graph is a finite sequence of vertices and edges of the form $v_1, e_1, v_2, e_2, v_3, e_3, \ldots, e_r, v_r$, where e_k is an edge between v_{k-1} and v_k. In general, the vertices and edges in a path need not be distinct.

A path is **simple** if its vertices are distinct. In a simple path, obviously all the edges are distinct. But a path with distinct edges can have repeated vertices. A graph is said to be **connected** if there is a path between every pair of vertices in it.

A path between a vertex and itself is a **closed path.** A closed path in which all the edges are distinct is a **circuit.** A circuit in which all the vertices are distinct is a **cycle.** Notice that v, e_1, w, e_2, v is a cycle but v, e, w, e, v is not a circuit and therefore not a cycle. These two closed paths can be represented as

$$v \; \text{--} \overset{e_1}{\text{-----}} \text{--} w \; \text{--} \overset{e_2}{\text{-----}} \text{--} v \qquad \text{and} \qquad v \; \text{-----} \overset{e}{\text{-----}} \text{--} w \; \text{-----} \overset{e}{\text{-----}} \text{--} v$$

In Figure 4.3.1, $1 \text{ ---- } 2 \text{ ---- } 3 \text{ ---- } 2 \text{ ---- } 1 \text{ ---- } 5$ is a path and $1 \text{ ---- } 2 \text{ ---- } 3 \text{ ---- } 4 \text{ ---- } 5$ is a simple path, and we can also see that $2 \text{ ---- } 3 \text{ ----- } 4 \text{ ---- } 5 \text{ ---- } 1 \text{ ---- } 2 \overset{e_1}{\text{--}} 4 \overset{e_2}{\text{--}} 2$ is a circuit and $2 \overset{e_1}{\text{-----}} 4 \text{ ----- } 3 \text{ ----- } 2$ is a cycle.

If v and w are connected (i.e., there is a path between them), then w and v are connected. In fact, the relation J defined by vJw if v and w are connected is an equivalence relation partitioning the set V of vertices into pairwise disjoint

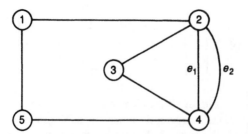

FIGURE 4.3.1

subsets of V. The subgraph induced by any such subset is a maximal connected subgraph called a **component** of the graph. The number of components of a graph G is denoted by $K(G)$ and is equal to 1 if and only if G is connected. The graph in Figure 4.3.2 has two components, G' and G'', where G' is induced by the subset $\{1, 2, 3, 4\}$ and G'' is induced by $\{5, 6, 7\}$.

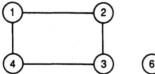 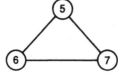

FIGURE 4.3.2

There is an interesting and useful relation between the number of paths between pairs of vertices in G and the elements of the powers of its adjacency matrix. A path with k edges is called a k-**path**. A 1-path is an edge. In the $n \times n$ adjacency matrix of the graph (with no multiple edges) with vertex set $V = \{1, 2, \ldots, n\}$, the (i, j)-element is 1 if and only if the number of 1-paths (edges) between i and j is 1. That this result can be generalized is the content of the following assertion.

THEOREM 4.3.1

If A is the adjacency matrix of a graph, the (i, j)-entry of the kth power $(k \geq 1)$ of A is the number of k-paths between vertex i and vertex j.

Proof:

The proof is by induction on k. This is true when $k = 1$. Assume that this is true for $(k - 1)$. Let $a_{ij}^{(r)}$ be the (i, j)-entry of the rth power of A. Then

$$a_{ij}^{(k)} = \sum_{p=1}^{p=n} a_{ip}^{(k-1)} a_{pj} \qquad \text{since } A^k = A^{k-1} \cdot A \qquad (*)$$

But

$$a_{ip}^{(k-1)} \, a_{pj} = \begin{cases} a_{ip}^{(k-1)} & \text{if } p \text{ and } j \text{ are adjacent} \\ 0 & \text{otherwise} \end{cases}$$

By hypothesis, $a_{ip}^{(k-1)}$ = number of $(k-1)$-paths between i and p, and this will be equal to the number of k-paths between from i to j, in which the vertex just prior to j is p. So the right side of (*) is the total number of k-paths between i and j obtained after examining $p = 1$ to $p = n$ consecutively.   ◇

COROLLARY

The (i, i) − entry in A^2 is the degree of i.

Example 4.3.1

As an illustration, consider the matrices A, $A,^2$ and A^4 of the graph of Figure 4.3.3. We have

$$A = \begin{bmatrix} 0 & 1 & 0 & 1 & 0 \\ 1 & 0 & 1 & 0 & 1 \\ 0 & 1 & 0 & 1 & 1 \\ 1 & 0 & 1 & 0 & 0 \\ 0 & 1 & 1 & 0 & 0 \end{bmatrix} \quad A^2 = \begin{bmatrix} 2 & 0 & 2 & 0 & 1 \\ 0 & 3 & 1 & 2 & 1 \\ 2 & 1 & 3 & 0 & 1 \\ 0 & 2 & 0 & 2 & 1 \\ 1 & 1 & 1 & 1 & 2 \end{bmatrix} \quad A^4 = \begin{bmatrix} 9 & 3 & 11 & 1 & 6 \\ 3 & 15 & 7 & 11 & 8 \\ 11 & 7 & 15 & 3 & 8 \\ 1 & 11 & 3 & 9 & 6 \\ 6 & 8 & 8 & 6 & 8 \end{bmatrix}$$

In A^2 the $(4, 4)$th entry is 2 and the degree of vertex 4 is 2 and the two 2-paths between 4 and 4 are 4 - - - - 1 - - - - 4 and 4 - - - - 3 - - - - 4. From the fourth power of A we see that there are eight different 4-paths between 2 and 5.

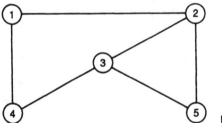

FIGURE 4.3.3

Finally, three more definitions: An edge in a connected graph is called a **bridge** if the removal of that edge, but not its end vertices, makes the graph disconnected. A graph with no cycles is an **acyclic graph**, also called a **forest**. A connected forest is a **tree.** We study trees in detail in Chapters 6 and 7.

4.4 *REACHING IN DIGRAPHS*

A **directed path** from a vertex v to a vertex w in a digraph is a finite sequence $v_1, a_1, v_2, a_2, \ldots, v_r, a_r, v_{r+1}$ of vertices and arcs, where the first vertex is v and the last vertex is w and a_i is an arc from v_i to v_{i+1}. If there is a directed path from v to w, then v is **connected to** w and w is **connected from** v. A pair of vertices is a **strongly connected pair** if each is connected to the other. If one of them is connected to the other, it is a **unilaterally connected pair.** A digraph is **strongly connected** if every pair of vertices is a strongly connected pair and it is **unilaterally connected** if every pair is unilaterally connected. A digraph is **weakly connected** if its underlying graph is connected.

A directed path from a vertex to itself is a **closed directed path.** A closed directed path is a **directed circuit** if its arcs are distinct and it is a **directed cycle** if its vertices are different. Notice the subtle difference between graphs and digraphs: If the vertices of a closed directed path are distinct, its arcs are distinct. But if the vertices of a closed path in a *graph* are distinct, its edges need not be distinct.

The relation defined by vRw if $\{v, w\}$ is a strongly connected pair is an equivalence relation giving a partition of the vertex set V into a class of pairwise disjoint subsets, and the subgraph induced by any one of these subsets is called a **strong component** of the digraph. For example, in the digraph of Figure 4.4.1 we have two strong components induced by the sets $\{1, 2, 3, 4\}$ and $\{5, 6, 7\}$.

As in the case of graphs the elements of the kth power of the adjacency matrix A of a digraph can be used to compute the number of k-paths between pairs of vertices. The proof of the following result is left as an exercise.

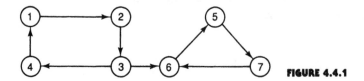

FIGURE 4.4.1

THEOREM 4.4.1

If A is the adjacency matrix of a digraph, then the (i, j)-entry of the kth power $(k \geq 1)$ of A is the number of k-directed paths from i to j.

If G is a graph, then a digraph G' obtained from G by changing each edge of G into an arc is called an **orientation** of G. For example, in Figure 4.4.2 the digraphs of (b) and (c) are both orientations of the graph of (a).

An orientation of a graph is called a **strong orientation** of the graph if the orientation is strongly connected. In Figure 4.4.2 the digraph in (c) is a strong orientation of the graph in (a). A graph is said to be **strongly orientable** if it

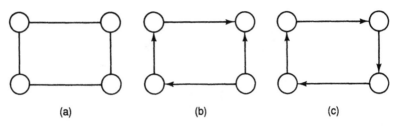

FIGURE 4.4.2

has a strong orientation. It is easy to see that a strongly orientable graph is necessarily connected and bridgeless. The converse also holds good: A **graph is strongly orientable if and only if it is connected and bridgeless.** This theorem is due to H. E. Robbins (1939) and we omit its proof. Later in the chapter we discuss an algorithm to obtain a strong orientation of a graph if such an orientation exists.

The **reachability matrix** of a digraph with n vertices is a $n \times n$ matrix $R = (r_{ij})$, where r_{ij} is 1 if there is a directed path from i to j, and 0 otherwise. Obviously, a digraph is strongly connected if and only if every element of its reachability matrix is equal to 1.

4.5 TESTING CONNECTEDNESS

Given a graph, it is natural to ask whether it is connected. Of course, from the diagram of a graph one can easily see whether the graph has more than one component and thereby test its connectedness. For large graphs such diagrams are not feasible. Moreover, if we input a graph into a computer, we need an algorithm to see whether it is connected. One such algorithm is the **depth-first search** (DFS) technique, in which we relabel the vertices of the graph as follows.

Let the vertices of the graph G be $v_1, v_2, \ldots,$ and v_n. Select an arbitrary vertex and label it as 1. Pick any vertex adjacent to 1. This is not yet labeled; label it as 2. Mark the edge $\{1, 2\}$ as a used edge so that it will not be used again. Proceeding similarly, suppose that we label vertex v_i with integer k. Search among all the unlabeled adjacent vertices of this vertex, select one of them, and label it as $(k + 1)$. Mark the edge $\{k, k + 1\}$ as a used edge. Now it may be the case that all the adjacent vertices of k are labeled. If so, go back to vertex $(k - 1)$ and search among its unlabeled adjacent vertices. If we find one such vertex, label it as $(k + 1)$ and mark the edge $\{k - 1, k + 1\}$ as a used edge. Continue the process until all the vertices are labeled or we are back at vertex 1 with at least one vertex unlabeled. In the former case the graph is connected and there will be exactly $(n - 1)$ used edges. The acyclic subgraph consisting of the n vertices of the graph and these $(n - 1)$ used edges is called a **depth-**

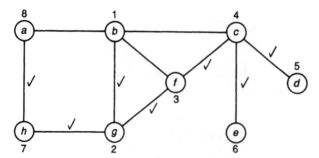

FIGURE 4.5.1

first search spanning tree of the graph. If it is not possible to label all the *n* vertices by the DFS technique, we conclude that the graph is not connected. A similar procedure for testing the strong connectedness of digraphs is given in Tarjan (1971).

Let us illustrate this DFS labeling technique to the graph of Figure 4.5.1 with eight vertices *a*, *b*, *c*, *d*, *e*, *f*, *g*, and *h*. We select vertex *b* and label it 1. An adjacent vertex to 1 is *g*. Label it as 2. Mark the edge {1, 2} as a used edge. A vertex adjacent to 2 and not labeled is *f*, and it is labeled 3. At this stage, edge {2, 3} is marked. We now see that *c* is the only unlabeled vertex adjacent to 3. So *c* is labeled as 4 and {3, 4} is marked. Then *d* or *e* can be labeled as 5. The tie is broken by labeling *d* as 5 and {4, 5} is marked. We notice that 5 has no unlabeled adjacent vertices, so we go back to 4 and label *e* as 6. At this stage we go back to 4, then to 3, and then to 2 in search of unlabeled adjacent vertices. We label *h* as 7 and *a* as 8 and mark the edges {2, 7} and {7, 8}. At this point all the eight vertices are labeled, showing that the graph is indeed connected. The seven marked edges are the edges of the DFS spanning tree as shown in Figure 4.5.2.

Now let us find the computational complexity of the DFS algorithm using the example just discussed. Notice that each edge {*i, j*}, where *i < j* can be investigated in the forward direction from *i* to *j* or in the backward direction from *j* to *i*. In our example we investigate {1, 2}, {2, 3}, {3, 4}, and {4, 5} in the forward direction. Since 5 had no unlabeled adjacent vertices, we had to go

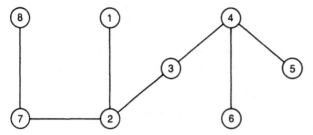

FIGURE 4.5.2

back to 4, which meant that we had to investigate $\{4, 5\}$ in the backward direction. So the edge $\{4, 5\}$ is examined twice and will never be investigated again. Thus each edge is examined at most twice. So if there are m edges, there will be at most $2m$ investigations, and there are n vertices to be labeled. Thus the complexity is at most $n + 2m$. Since the maximum value for m is $n(n - 1)/2$, the worst-case complexity of the DFS algorithm is n^2.

4.6 STRONG ORIENTATION OF GRAPHS

Consider a graph model in which the vertices are the street corners of a large city. Two vertices are joined by an edge if there is a street joining them. Suppose that the resulting graph G is strongly orientable. This is the case if and only if G is connected and bridgeless. We are now interested in (temporarily) converting all streets in the city into one-way streets. Since G is strongly orientable every corner can be reached from every other corner after this conversion. How is this conversion carried out? We again resort to the DFS procedure and label all the vertices. If $\{i, j\}$ is a marked edge where $i < j$, convert this edge into an arc from i to j. On the other hand, if $\{i, j\}$ is an unmarked edge where $i < j$, convert this edge into an arc from j to i. The resulting digraph G' is a strong orientation of G. For a proof of this assertion, see Roberts (1976). In Figure 4.6.1 we have a connected bridgeless graph in (a), with a DFS spanning tree in (b), where the vertices are appropriately labeled. A strong orientation of G is the digraph G' of Figure 4.6.1(c).

When we use the DFS procedure there will be m unmarked edges in the worst case that have to be investigated. Thus the complexity is at most $n + 2m + m$, which in the worst case will be equal to $f(n) = (3n^2 - n)/2$.

4.7 NOTES AND REFERENCES

There are several excellent references on graph theory both at the introductory level and at the advanced level. Here is a partial list: Behzad et al. (1979), Berge (1962), Bondy and Murty (1976), Carre (1979), Chartrand (1977), Deo (1974), Gibbons (1985), Gondran and Minoux (1984), Harary (1969a), Ore (1963), Roberts (1976, 1978), Swamy and Thulasiraman (1981), Wilson (1979), and Yemelichev, et al. (1984). The chapters on graphs in the books by Grimaldi (1985), Liu (1985), Roberts (1984), Townsend (1987), and Tucker (1984) are also highly recommended. For a proof of Robbins's theorem, see Chapter 2 of Roberts (1978), which also contains a a complete discussion of strong orientations and one-way street assignments.

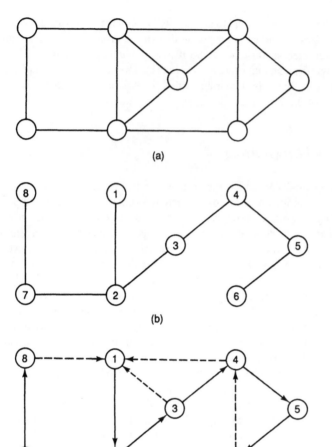

(a)

(b)

(c)

FIGURE 4.6.1

4.8 EXERCISES

4.1. Draw a graph $G = (V, E)$, where $V = \{1, 2, 3, 4, 5\}$ and $E = \{\{1, 2\}, \{1, 3\}, \{1, 5\}, \{2, 3\}, \{3, 4\}, \{3, 5\}, \{4, 5\}\}$. Find the set $W = \{i : i$ is a vertex such that i and 2 are adjacent$\}$.

4.2. Draw a digraph for which the underlying graph is the graph of Problem 4.1. Find the set of vertices **(a)** adjacent to vertex 2 and **(b)** adjacent from vertex 2 in this digraph.

4.3. **(a)** Construct a complete graph with four vertices such that no two edges intersect. **(b)** Construct a complete graph with five vertices. **(c)** Do you notice any difference between the graphs in parts **(a)** and **(b)**?

4.4. Find the number of edges in a complete graph with n vertices.

4.5. Draw an air transportation graph with Boston, New York, London, Paris, Moscow, Prague, and Rome as vertices and an edge joining two cities if there is nonstop air service between them.

4.6. Draw the subgraph induced by $V' = \{2, 3, 4, 5\}$ in Problem 4.1.

4.7. Find the minimum number of bridges (a) to be constructed and (b) to be demolished so that the Königsberg bridge problem becomes solvable.

4.8. (a) Draw the bipartite graph $K_{2,2}$ such that no two edges intersect. (b) Draw the bipartite graph $K_{3,3}$. (c) What is the noticeable difference between these two bipartite graphs?

4.9. Find the number of edges in $K_{p,q}$.

4.10. Consider the graph $G = (V, E)$ with $V = \{1, 2, 3, 4, 5\}$ and $E = \{a, b, c, d, e, f, g, h\}$, where $a = \{1, 2\}$, $b = \{2, 3\}$, $c = \{3, 5\}$, $d = \{2, 5\}$, $e = \{2, 4\}$, $f = \{4, 5\}$, $g = \{1, 4\}$, and $h = \{1, 5\}$.
   (a) Find the adjacency matrix of G.
   (b) Find the degree of each vertex and find the set of of odd vertices.
   (c) Find the incidence matrix of G.

4.11. Consider a digraph G' the underlying graph of which is the graph G of Problem 4.1.
   (a) Find the adjacency matrix of G'.
   (b) Find the indegree and undegree of each vertex in G'.
   (c) Find the incidence matrix of G'.

4.12. Give an example of a simple graph with (a) no odd vertices, (b) with no even vertices.

4.13. Construct a connected simple graph with n vertices such that the degree of each vertex is 2. Notice the structure of the graph.

4.14. Construct a graph with n vertices and $(n - 1)$ edges such that there are two vertices of degree 1 and $(n - 2)$ vertices of degree 2.

4.15. A graph G with the property that all of its vertices have the same degree r is called a **regular graph** of degree r. Notice that a complete graph is regular but the converse is not true. (a) Construct a simple regular graph of degree 1 that is not complete. (b) Construct a simple regular graph of degree 2 that is not complete. (c) If G is a regular graph of degree r and if G has n vertices, find the number of edges in G.

4.16. Prove that in a simple graph with two or more vertices, the degrees of the vertices cannot all be distinct.

4.17. Let G be a simple graph with n vertices with A as its adjacency matrix and B as its incidence matrix. Define the $n \times n$ diagonal matrix C in which the ith diagonal element is the degree of vertex i in G. C is called the **degree matrix** of G. Prove that $B \cdot B^t = A + C$.

4.18. A square matrix in which each element is 0 or 1 is called a **dominance matrix** if (a) each diagonal number is 0 and (b) the (i, j) element is 1 if and only if the

(j, i) element is 0. Prove that the adjacency matrix of a tournament is a dominance matrix.

4.19. Consider the digraph $G = (V, E)$ where $V = \{1, 2, 3, 4, 5, 6\}$ and $E = \{(1, 2), (2, 3), (3, 4), (4, 5), (5, 6), (1, 6), (2, 6), (5, 2)\}$.

(a) Find a 6-path from 1 to 6.

(b) Find a simple path from 1 to 6 using five arcs.

(c) Find a cycle with four arcs.

(d) Use the adjacency matrix of G to determine the number of 2-paths from 2 to 4.

(e) Find all the strong components of G.

(f) Find the reachability matrix R of G.

4.20. Let R be the reachability matrix of a digraph $G = (V, E)$ where $V = \{1, 2, \ldots, n\}$, $P = (p_{ij})$ be the *element-wise product* of R and the transpose of R, and $Q = R^2 = (q_{ij})$. Prove:

(a) $p_{ij} = 1$ if and only if i and j are strongly connected

(b) $q_{ii} =$ the number of vertices in the strong component which contains vertex i.

4.21. Construct a graph with five vertices and six edges that consists of a circuit with six edges and two cycles with three edges each.

4.22. Consider the graph of Problem 4.1.

(a) Find a 6-path between 1 and 4.

(b) Find a simple path between 1 and 4 with four edges.

(c) Use the adjacency matrix to determine the number of 2-paths between 2 and 4.

4.23. Define the reachability matrix of a *graph*.

4.24. Draw graph G with adjacency matrix A such that

$$A^2 = \begin{bmatrix} 1 & 0 & 1 & 1 \\ 0 & 3 & 1 & 1 \\ 1 & 1 & 2 & 1 \\ 1 & 1 & 1 & 2 \end{bmatrix} \quad \text{and} \quad A^3 = \begin{bmatrix} 0 & 3 & 1 & 1 \\ 3 & 2 & 4 & 4 \\ 1 & 4 & 2 & 3 \\ 1 & 4 & 3 & 2 \end{bmatrix}$$

4.25. Show that the sum of the diagonal elements of the second power of the adjacency matrix of a graph G is twice the number of edges in G.

4.26. If G is a connected graph with n vertices, show that there exists a path with at most $(n - 1)$ edges between every pair of vertices.

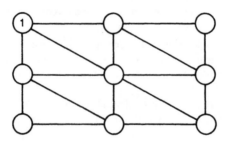

FIGURE 4.0.1

4.27. If A is the adjacency matrix of a graph with n vertices and if a nondiagonal element of $A + A^2 + A^3 + \cdots + A^{n-1}$ is zero, what can you say about G?

4.28. Obtain a DFS spanning tree (starting from vertex 1) in the connected graph G of Figure 4.8.1.

4.29. The graph G in Problem 4.28 is connected and bridgeless. Find a strong orientation of G.

4.30. A triangle is said to be monochromatic if all its sides are of the same color. Show that no matter how we color the edges of a complete graph with six vertices using two colors, there will always be at least one monochromatic triangle. (See Example 1.5.5.) It can be shown that there will be at least two such triangles. Show also that it is possible to color all the edges of a complete graph with five vertices using two colors such that there is no monochromatic triangle.

CHAPTER

5

More on Graphs and Digraphs

5.1 EULERIAN PATHS AND EULERIAN CIRCUITS

A path in a graph is an **Eulerian path** if every edge of the graph appears as an edge in the path exactly once. A closed Eulerian path is an **Eulerian circuit.** A graph is said to be an **Eulerian graph** if it has an Eulerian circuit. There are analogous definitions in the case of digraphs.

The idea of Eulerian circuits first arose from the famous Königsberg bridge problem (Example 4.1.1), which asked whether one could traverse all the seven bridges in the town, going over each one exactly once, and returning to the starting location. In the course of demonstrating that it was impossible to do so, Euler produced techniques which, it is universally believed, gave birth to graph theory. It is obvious that the problem can be solved if its graph model (see Figure 4.1.6) is an Eulerian graph. The following theorem settled this question.

THEOREM 5.1.1

A connected graph G with no loops is Eulerian if and only if the degree of each vertex is even.

Proof:

Any Eulerian circuit in G leaves each vertex as many times as it enters. So each vertex of G is even. On the other hand, suppose that G is a connected graph in which each vertex is even. We prove that G is Eulerian by actually constructing an Eulerian circuit in it. There are several algorithms for this construction. For details, refer to Even (1979). We adopt the following procedure, in which circuits are "spliced," until we actually obtain an Eulerian circuit. Start from any vertex v. Traverse distinct edges of G until we return to v. This is certainly possible since each vertex is even. Let C_1 be the circuit thus obtained. If this circuit contains all the edges of the graph, we are done. Otherwise, delete all the edges of this circuit and all vertices of degree 0 from G to obtain the connected subgraph H_1 in which each vertex is also even. Furthermore, since G is connected, there is a vertex u common to both the circuit C_1 and the subgraph H_1. Now start from u and obtain a circuit C_2 by traversing distinct edges of the subgraph. Notice that the two circuits have no common *edges*, even though they may have common vertices. If $v = u$, the two circuits can be joined together to form an enlarged circuit C_3. See Figure 5.1.1(a). If v and u are distinct, let P and Q be the two distinct simple paths between v and u consisting of edges from C_1. Then P, Q, and C_2 are spliced together to form a new circuit C_3, as in Figure 5.1.1(b). If this enlarged circuit has all the edges of G, we conclude that it is Eulerian. Otherwise, we continue until we obtain a circuit that has all the edges of G.   ◇

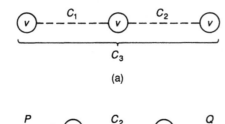

(a)

(b)

FIGURE 5.1.1

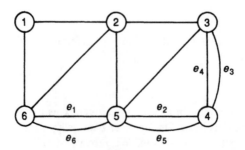

FIGURE 5.1.2

To illustrate this procedure, let us try to construct an Eulerian circuit for the graph G of Figure 5.1.2 in which the edges are labeled whenever there are multiple edges. Starting from vertex 1, suppose that we have the circuit C_1, consisting of $\{1, 2\}$, $\{2, 3\}$, e_3, e_2, e_1, and $\{6, 1\}$. Deleting all the edges of this circuit from G and then deleting all vertices of degree zero, we get the subgraph H_1 as in Figure 5.1.3 and we see that vertex 3 is common for both the subgraph and the circuit C_1. Starting from vertex 3 in this subgraph, we get the circuit C_2 consisting of e_4, e_5, and $\{5, 3\}$. Then we splice these two circuits to get a circuit C_3 consisting of $\{1, 2\}$, $\{2, 3\}$, all the edges of C_2, e_3, e_2, e_1, and $\{6, 1\}$. This new circuit also is not Eulerian, leaving us with a new subgraph H_2 as in Figure 5.1.4. The vertex 2 is common for this subgraph and C_3 and we have a circuit C_4 in H_2 consisting of $\{2, 5\}$, e_6, and $\{6, 2\}$. Finally, we splice C_4 and C_3 to obtain an Eulerian circuit of G consisting of $\{1, 2\}$, $\{2, 5\}$, e_6, $\{6, 2\}$, $\{2, 3\}$, e_4, e_5, $\{5, 3\}$, e_3, e_2, e_1, and $\{6, 1\}$.

Finding an Eulerian circuit by this method can be tedious, particularly in large graphs. The following procedure, known as Fleury's algorithm, is less complicated: Start from any vertex and delete an edge as soon as it is traversed. Also, never cross a bridge if you can help it. If we are able to return to the starting point after deleting all the edges, the circuit is Eulerian and we conclude that the graph is Eulerian as well. For example, in Figure 5.1.2 we start from 2 and traverse along $\{2, 3\}$, $\{3, 5\}$, $\{5, 2\}$, $\{2, 1\}$, and $\{1, 6\}$ and stop at 6. If we delete all the edges traversed, we get the subgraph as in Figure 5.1.5, in which we start from 6 but we do not go along e_7 since it is a bridge. So we

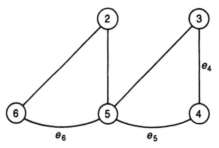

FIGURE 5.1.3

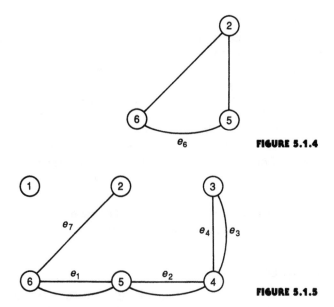

FIGURE 5.1.4

FIGURE 5.1.5

traverse along e_1, e_2, and e_3, reaching 3. Once these edges are deleted, e_4 becomes a bridge that we are forced to cross, and similarly, we cross the bridges e_5, e_6 and finally $\{6, 2\}$. At this stage we have an Eulerian circuit.

Note: The presence of a loop at a vertex does not in any way influence the existence of an Eulerian circuit. Let G be any connected graph and let G' be the subgraph obtained after deleting all its loops. Then G is Eulerian if and only if G' is Eulerian. Unless otherwise stated, we assume that all graphs and digraphs in the remainder of this chapter are loopless.

A characterization of graphs with Eulerian paths can now easily be obtained as follows.

THEOREM 5.1.2

A connected non-Eulerian graph G with no loops has an Eulerian path if and only if it has exactly two odd vertices.

Proof:

If G has an Eulerian path from u to v, both u and v are odd and since this path passes through every vertex and traverses each edge once, every other vertex is necessarily even. On the other hand, suppose that G is connected with exactly two odd vertices, u and v. Now either u and v are adjacent or they are not. In the former case let e be an edge between them. Delete e to get the graph G' (with at most two components) in which each vertex is even. If G' is connected, obtain an Eulerian circuit in it starting from u and then adjoin the edge e to this circuit to get an Eulerian path between u and v. If G' has two components, let the component that contains u be G_1 and the component that contains v be G_2. Of course, both these components are Eulerian. Now obtain an Eulerian circuit from u in the first component and an Eulerian circuit from v in the second component. Then the path consisting of the edges of the first circuit, the edge (actually, the bridge) e, and the edges of the second circuit constitute an Eulerian path between u and v. Finally, if u and v are not adjacent in G, construct an arc e joining them, producing a new graph H, which is Eulerian. Obtain an Eulerian circuit in H from u in which the last edge is e. If we delete e, we have an Eulerian path in G from u to v. This completes the proof. $\diamondsuit$

We now state analogous results in the case of digraphs. For the straightforward proofs of these theorems, the reader is referred to Behzad et al. (1979).

THEOREM 5.1.3

A weakly connected digraph has a directed Eulerian circuit if and only if the indegree of each vertex equals its outdegree.

THEOREM 5.1.4

A weakly connected digraph with no directed Eulerian circuit has a directed Eulerian path if and only if the indegree of each vertex equals its outdegree except for two vertices u and v such that the outdegree of u equals its indegree plus one and the indegree of v equals its outdegree plus one.

5.2 CODING AND DE BRUIJN DIGRAPHS

There are several interesting and useful applications of Eulerian paths and circuits in many areas, such as computer science, operations research, cryptography, and transportation problems, to name a few. We discuss some examples here. The Chinese postman problem is a network optimization problem in which an arbitrary connected network is enlarged into an Eulerian one. The problem can be stated as follows: A mail carrier starts from the post office, delivers mail to each block in his beat, and returns to the post office. If we take each street corner in the route as a vertex and a street between two corners as an edge, we have a graph G as a model of this problem. If G is Eulerian, the mail carrier has to traverse each street exactly once. If it is not Eulerian, he has to repeat some edges. A typical optimization problem in this context is to locate those streets which have to be repeated so that the total distance traversed is a minimum.

This was first discussed by the Chinese mathematician Kwan (1962) and so is known as the Chinese postman problem. In this section we discuss an application of Eulerian graphs to coding theory.

Any word with m letters out of which n are distinct can be associated with a weakly connected digraph G with n vertices and $m - 1$ arcs such that the word represents a directed Eulerian path if the first letter and last letter are different and a directed Eulerian circuit if the first letter and the last letter are the same. For example, in the word "LETTERED" we have $m = 8$ and $n = 5$, and this letter can be associated with the directed path from the vertex L to the vertex D in the digraph of Figure 5.2.1 with five vertices and seven arcs represented by L- - -E- - -T- - -T- - -E- - -R- - -E- - -D, which is a directed Eulerian path. Similarly, the word "ELECTIVE" can be associated with a directed Eulerian cycle E- - -L- - -E- - -C- - -T- - -I- - -V- - -E in the digraph in Figure 5.2.2. Notice that even though a word defines a digraph uniquely, it is possible that the same digraph can be associated with several words of equal length.

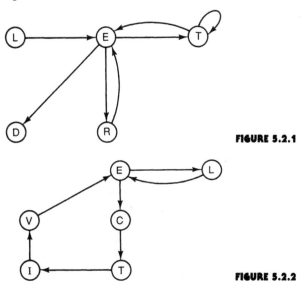

FIGURE 5.2.1

FIGURE 5.2.2

In any word with n distinct letters $A_1, A_2, \ldots, A_n$, let $f(A_i)$ be the frequency of the letter A_i in the word. Then the sum of the frequencies of the n letters is m. Let m_{ij} denote the number of times A_j appears *immediately* after A_i, denoting the number of arcs from A_i to A_j in the digraph. For example, in the word "MATHEMATICS," $m_{AT} = 2$, $m_{TA} = 0$, $m_{TH} = 1$, and so on. Let $M = (m_{ij})$ be the $n \times n$ matrix thus defined. The row sum of row i in M is the outdegree of vertex A_i, and the column sum of column j is the indegree of A_j.

Thus, corresponding to each word with n distinct letters, we have a frequency set of n positive integers and a $n \times n$ matrix whose elements are nonnegative integers. For example, in the word "LETTERED" the five distinct letters are D, E, L, R, and T, with a frequency set $\{1, 3, 1, 1, 2\}$, and the 5×5 matrix is

	D	E	L	R	T	Row sum
D	0	0	0	0	0	0
E	1	0	0	1	1	3
$M =$ L	0	1	0	0	0	1
R	0	1	0	0	0	1
T	0	1	0	0	1	2
Column sum:	1	3	0	1	2	

In the word "ELECTIVE" the six distinct letters are C, E, I, L, T, and V, with a frequency set $\{1, 3, 1, 1, 1, 1\}$, and the 6×6 matrix is

	C	E	I	L	T	V	Row sum
C	0	0	0	0	1	0	1
E	1	0	0	1	0	0	2
I	0	0	0	0	0	1	1
$M =$ L	0	1	0	0	0	0	1
T	0	0	1	0	0	0	1
V	0	1	0	0	0	0	1
Column sum:	1	2	1	1	1	1	

We now make the following easily verifiable assertions:

1. The digraph of any word is weakly connected.
2. If the first letter and the last letter are not the same, the row sum of the first letter equals the column sum of the first letter plus one, the row sum of the last letter equals the column sum of the last letter minus one, and for all other letters the row sum and the column sum are equal.
3. If the first letter and the last letter are the same, then row sum equals the column sum for all letters and the row sum of the starting letter will be one less than its frequency.

All the information in a codeword is contained in the frequency set and the matrix M associated with the word. Hutchinson and Wilf (1975) study codewords in their investigation of DNA and RNA molecules. Suppose that we are given (a) a set of n letters $A_1, A_2, \ldots, A_n$, (b) a set of n positive integers f_1,

$f_2, \ldots, f_n$, and (c) a $n \times n$ matrix (m_{ij}) of nonnegative integers. Does there exist a word in which A_i appears exactly f_i times and A_j appears immediately after A_i exactly m_{ij} times? The answer is "yes" if conditions corresponding to assertions (1)–(3) are satisfied because of Theorems 5.1.3 and 5.1.4. We thus have the following result.

THEOREM 5.2.1

Let $M = (m_{ij})$ be a $n \times n$ matrix with nonnegative integer components and let A_i $(i = 1, 2, \ldots, n)$ be a set of n distinct letters such that A_i is associated with both row i and column i. Let

$$r_i = \text{sum of all the elements of row } i \text{ of } M$$
$$c_j = \text{sum of all the elements of column } j \text{ of } M$$

(a) If $r_j = c_j + 1$, $r_k = c_k - 1$, where j and k are distinct and if $r_i = c_i$ in all other cases, there exists a word beginning with A_j and ending in A_k in which the frequency of A_j is r_j, the frequency of A_k is c_k, and the frequency of every other letter is r_i, which is also c_i. Moreover, in the word, the letter A_p appears immediately after A_q exactly m_{pq} times.

(b) If $r_i = c_i$ for $i = 1, 2, \ldots, n$, and if f_i $(i = 1, 2, \ldots, n)$ are nonnegative integers such that $r_j = f_j - 1$ and $r_i = f_i$ for every i other than j, there exists a word that begins with A_j and ends with A_j in which A_k appears exactly f_k times and A_p appears after A_q exactly m_{pq} times.

For example, suppose that the distinct letters in a word are A, B, C, and D and the matrix is

	A	B	C	D	Row sum
A	0	1	1	0	2
B	1	0	1	0	2
C	0	0	0	2	2
D	1	0	0	0	1
Column sum:	2	1	2	2	

where $M =$ is to the left of the matrix.

First we construct a digraph G with four vertices A, B, C, and D as in Figure 5.2.3. We observe:

1. The digraph is weakly connected.
2. (Row sum for B) = (column sum for B) + 1.
3. (Row sum for D) = (column sum for D) − 1.
4. Row sum = column sum for all other letters.

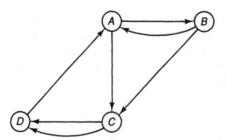

FIGURE 5.2.3

So there is a word that starts with B and ends in D which can be represented
by an Eulerian path from B to D in the digraph of Figure 5.2.3. One such word
is BABCDACD. It is an easy exercise to show that if the frequency set is
$\{4, 3, 5, 2\}$ and the matrix is

$$
M = \begin{array}{c@{\;}c}
 & \begin{array}{cccc} A & B & C & D \end{array} \\
\begin{array}{c} A \\ B \\ C \\ D \end{array} &
\left[\begin{array}{cccc}
1 & 0 & 2 & 1 \\
1 & 0 & 1 & 1 \\
1 & 2 & 1 & 0 \\
1 & 1 & 0 & 0
\end{array}\right]
\end{array}
$$

then a word is CCBCBAADBDACAC.

   We conclude this chapter with a discussion of de Bruijn digraphs, another
application of Eulerian digraphs. There are 2^{n-1} binary words of length $n - 1$.
We construct a digraph with 2^{n-1} vertices as follows. Let each word of length
$n - 1$ be a vertex. From each vertex of the form $v = a_1 a_2 \cdots a_{n-1}$ draw two
arcs: one to $a_2 a_3 \cdots a_{n-1}0$ and the other to $a_2 a_3 \cdots a_{n-1}1$ to represent two
n-letter words $v0$ and $v1$, respectively. So the 2^n arcs of the digraph thus constructed represent the set of binary words of length n. This digraph $G(2, n)$
known as the **de Bruijn digraph,** is weakly connected and is Eulerian since the
indegree of each vertex equals its outdegree. The digraph $G(2, 3)$ is as shown
in Figure 5.2.4.

   More generally, for an alphabet of p letters, $G(p, n)$ is a de Bruijn digraph
with p^{n-1} vertices and p^n arcs such that the indegree and outdegree of each vertex
are both p. Thus $G(p, n)$ is Eulerian. Now consider any Eulerian circuit in this
digraph that will contain all the p^n arcs in a sequence. Construct the sequence
of the first letters of all these words. Let us denote this sequence by $a_1 a_2 \cdots$
a_r, where $r = p^n$. Then the r distinct words of length n are all of the form
$a_i a_{i+1} \cdots a_{i+n-1}$, where the addition operation defined on the subscript is
modulo r. For example, if $p = 2$ and $n = 3$, then a_9 is a_1. In the digraph of
Figure 5.2.4 a directed Eulerian circuit starting from 00 consists of the following
sequence of eight arcs: 000, 001, 011, 111, 110, 101, 010, 100. The first letters

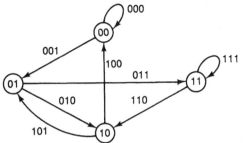

FIGURE 5.2.4

of these arcs form the word 00011101, so that $a_1 = 0$, $a_2 = 0$, $a_3 = 0$, $a_4 = 1$, $a_5 = 1$, $a_6 = 1$, $a_7 = 0$, and $a_8 = 1$. Any three-letter word is now of the form $a_i a_{i+1} a_{i+2}$. Thus $a_7 a_8 a_9 = a_7 a_8 a_1 = 010$, and so on.

We can formally define a de Bruijn sequence for two positive integers p and n. If S is any alphabet consisting of p letters, then a sequence $a_1 a_2 \cdots a_r$ of r ($r = p^n$) letters is called a **de Bruijn sequence,** denoted by $B(p, n)$, if every word of length n from S can be realized as $a_i a_{i+1} \cdots a_{i+n-1}$ ($i = 1, 2, \ldots, r$), where the addition operation in the subscripts is modulo r.

We are now ready to summarize our observations as a theorem.

THEOREM 5.2.2

For every pair of positive integers there exists a de Bruijn sequence.

This was first proved by de Bruijn (1946) for $p = 2$ and later generalized for arbitrary p by Good (1946). These sequences are very useful in coding theory. The state diagram of a feedback shift register (FSR) is a subgraph of a certain de Bruijn digraph. FSRs have a wide range of applications in communications, cryptography, and computer science, particularly so because of the randomness properties of the sequences they generate. Briefly speaking, if K is a field (of order q), and if $f: K^n \rightarrow K$, then an n-stage FSR on K transforms the vector $[x_0 \quad x_1 \quad \cdots \quad x_{n-1}]$ into the vector $[x_1 \quad x_2 \quad \cdots \quad x_n]$, where $x_n = f(x_0, x_1, \ldots, x_{n-1})$.

5.3 HAMILTONIAN PATHS AND HAMILTONIAN CYCLES

A path between two vertices in a graph is a **Hamiltonian path** if it passes through each vertex exactly once. A closed path that passes through each vertex exactly once and in which all the edges are distinct is a **Hamiltonian cycle.** A graph is a **Hamiltonian graph** if it has a Hamiltonian cycle. In a digraph a directed path from a vertex to another vertex is a **directed Hamiltonian path** if it passes

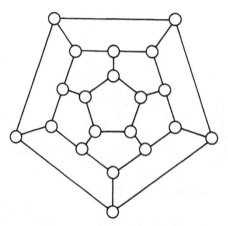

FIGURE 5.3.1

through each vertex exactly once. A closed directed Hamiltonian path is a **directed Hamiltonian cycle.** The adjective "Hamiltonian" is in honor of the famous Irish mathematician Sir William Hamilton (1805–1865), who investigated the existence of a solution to a game called "all around the world," in which the player is asked to find a route along the edges of a dodecahedron (a regular polyhedron with 20 vertices, 30 edges, and 12 faces) visiting each vertex exactly once and returning to the starting vertex. Now a dodecahedron can be represented as a graph G on the plane (see Figure 5.3.1) with 20 vertices and 30 edges. Thus the game has a solution if and only if G is a Hamiltonian graph.

Even though the problem of determining the existence of Hamiltonian cycles appears similar to that of determining the existence of Eulerian circuits, it is not at all easy to tell whether a given graph is Hamiltonian in general. In contrast with the extremely tidy necessary and sufficient conditions obtained by Euler for the existence of Eulerian circuits, Hamiltonian graphs seem to defy characterization. In many cases each graph must be considered individually since no easily verified necessary conditions are known in general. Of course, a complete graph is Hamiltonian. In other words, a graph with n vertices is Hamiltonian if the degree of each vertex is $n - 1$. The larger the degree of each vertex, the more likely it appears that the graph is Hamiltonian. So the question is this: Does there exist a positive integer k ($k < n - 1$) such that the graph is Hamiltonian whenever the degree of each vertex is at least k? The answer is yes, as proved by Dirac (1952), whose theorem can also be obtained as a consequence of the following theorem of Ore (1963).

THEOREM 5.3.1

A simple graph with n vertices (where n is at least 3) is Hamiltonian if the sum of the degrees of every pair of nonadjacent vertices is at least n.

Proof:

Suppose that a graph G with n vertices is not Hamiltonian. So it is a subgraph of the complete graph K_n with fewer edges. Now keep on adding edges to G by joining nonadjacent vertices until we obtain a non-Hamiltonian graph H such that the addition of *one more edge* to H will make it Hamiltonian. Let x and y be any pair of nonadjacent vertices in H. So they are nonadjacent in G as well. Thus (deg x + deg y) is at least n in H. Since the addition of $\{x, y\}$ as an edge to H will make it Hamiltonian, there is a Hamiltonian path in H between x and y. If we write $x = v_1$ and $y = v_n$, then this path can be written as $v_1 \text{- - -} v_2 \text{- - -} v_3 \ldots v_{i-1} \text{- - -} v_i \text{- - -} v_{i+1} \ldots v_{n-1} \text{- - -} v_n$. Notice that if v_1 and v_i are adjacent in H, then v_n and v_{i-1} cannot be adjacent because if they are adjacent, we will have the following Hamiltonian cycle in H: $v_n \text{- - -} v_{i-1} \ldots v_1 \text{- - -} v_i \ldots v_n$, which is a contradiction. So if v_1 has r adjacent vertices from the set $\{v_2, v_3, \ldots, v_n\}$, at least r vertices from the set $\{v_1, v_2, \ldots, v_{n-1}\}$ cannot be adjacent to v_n. In that case, deg $v_1 = r$ and deg $v_n \leq (n-1) - r$ and consequently, deg v_1 + deg $v_n \leq (n-1) < n$, which contradicts the hypothesis. $\diamondsuit$

COROLLARY (Dirac's Theorem)

A simple graph with n vertices is Hamiltonian if the degree of each vertex is at least $n/2$.

Note: The converse of Ore's theorem is not true. For example, consider the graph of a polygon with six sides.

A sufficient condition for the existence of a Hamiltonian Path in a graph is as in the following result.

THEOREM 5.3.2

A simple graph with n vertices has a Hamiltonian path if the sum of the degrees of every pair of nonadjacent vertices is at least $(n-1)$.

Proof:

This is an exercise. $\diamondsuit$

COROLLARY

A simple graph with n vertices has a Hamiltonian path if the degree of each vertex is at least $(n-1)/2$.

Just as in the case of graphs, there is no known characterization of Hamiltonian digraphs. In fact, the situation becomes more complex. We state here a few sufficient conditions for the existence of direct Hamiltonian cycles and paths in simple digraphs which are more or less similar to the results in the case of graphs. See Behzad et al. (1979) for proofs.

THEOREM 5.3.3

(a) A strongly connected digraph with n vertices is a Hamiltonian digraph if (deg u + deg v) is at least $2n - 1$ for every pair of vertices u and v such that there is no arc from u to v and from v to u.

(b) A digraph with n vertices is Hamiltonian if (outdegree u + indegree v) is at least n for every pair of vertices u and v such that there is no arc from u to v.

(c) A strongly connected digraph with n vertices is Hamiltonian if (outdegree v + indegree v) is at least n for every vertex v.

(d) A digraph with n vertices is Hamiltonian if both the outdegree and indegree of each vertex is at least $n/2$.

THEOREM 5.3.4

(a) If (degree u + degree v) is at least $2n - 3$ for every pair of vertices u and v such that there is no arc from one to the other in a digraph G, then G has a directed Hamiltonian path.

(b) If (outdegree u + indegree v) is at least $(n - 1)$ for every pair of vertices such that there is no arc from u to v in a digraph G, then G has a directed Hamiltonian path.

(c) If (outdegree v + indegree v) is at least $(n - 1)$ for every vertex v in a digraph with n vertices, the digraph has a directed Hamiltonian path.

(d) If both the outdegree and indegree of each vertex is at least $(n - 1)/2$ in a digraph G, then G has a directed Hamiltonian path.

Hamiltonian-Connected Graphs

A graph is said to be **Hamiltonian-connected** if there is a Hamiltonian path between every pair of vertices in it. Obviously any Hamiltonian-connected graph with three or more vertices is necessarily a Hamiltonian graph. The converse is not true: In the Hamiltonian graph $G = (V, E)$ where $V = \{1, 2, 3, 4\}$ and E is the set $\{\{1, 2\}, \{2, 3\}, \{3, 4\}, \{4, 1\}\}$ there is no Hamiltonian path between the vertex 2 and the vertex 4. The following result due to Ore (1963) parallels the result of Theorem 5.3.1.

THEOREM 5.3.5

If G is a simple graph with n vertices (n is at least 3) such that for all distinct nonadjacent vertices i and j, (degree of i) + (degree of j) exceeds n, then G is Hamiltonian-connected.

For more details on Hamiltonian-connected graphs, see the papers by Chartrand et al. (1969) and Lick (1970).

5.4 APPLICATIONS OF HAMILTONIAN CYCLES

Hamiltonian paths and cycles have several useful and interesting applications. We discuss some of them here.

Example 5.4.1 (The Traveling Salesman Problem)

In Example 4.1.3 we introduced the traveling salesman problem (TSP), in which a salesman has to make an itinerary visiting each city on the tour exactly once and returning to the starting point. Any such tour is a Hamiltonian cycle. Assuming that such a cycle exists, the optimization problem then is to find such a tour for which the total cost (or for that matter, total distance) is a minimum. TSP is one of the best known problems of a class of those that are easy to state but very difficult to solve. In general, there is no known efficient procedure for finding a solution to the problem. If we adopt the exhaustive enumeration method in which we list all the $(n - 1)!$ directed Hamiltonian cycles (in the worst case) is a digraph with n vertices, and compute the cost for each cycle by performing the n additions, we will be doing $n(n - 1)!$ additions. If $n = 20$ and a computer can do 1 million additions per second, this method will take about 75,000 years. See Held and Karp (1970) for a discussion of the complexity considerations involved in this problem.

Example 5.4.2 (Scheduling)

Consider a machine shop with n different machines. A job has to be run through all these machines but not in any particular order. Let each machine represent a vertex of a digraph. Draw an arc from each vertex to every other vertex. Then any directed Hamiltonian path in the digraph is a schedule. If c_{ij} is the setup time required whenever the job goes from machine i to machine j, the optimization problem is to find a schedule that takes the least amount of time.

Example 5.4.3

In Example 4.1.4 we defined a tournament as a simple digraph in which for every pair of vertices v and w either there is an arc from v to w or from w to v but not both. In other words, the underlying graph of a tournament is complete and (v, w) is an arc in the tournament if and only if (w, v) is not an arc. If the vertices denote the different players, the existence of an arc from v to w indicates that v defeats w in the game. The following questions arise: (1) Is it possible to rank all the players as a sequence u_1, $u_2, \ldots, u_n$ such that u_i defeats u_{i+1} ($i = 1, 2, \ldots, n - 1$)? In other words, does the digraph have a directed Hamiltonian path? (2) If such a path exists, is it unique? (3) Is there a necessary and sufficient condition to be satisfied by the digraph so that the path is unique, implying that the ranking is also unique? The answers are: (1) yes, (2) no, and (3) yes.

Before we justify these assertions, let us consider the two tournaments (a) and (b), as in the digraphs of Figure 5.4.1. We see that both the tournaments have Hamiltonian paths. In (a) we have four different Hamiltonian paths, whereas in (b) we have only one. What makes these two different? Here is a definition to resolve this. A tournament is said to be **transitive** if whenever u defeats v and v defeats w, then u defeats w. Notice that a tournament is transitive if and only if it has no directed cycles with three arcs. Even though "transitivity" appears to be normal in most situations in life, in the real world most tournaments are not transitive. Maybe this is one reason they are exciting. In our illustration, (b) is transitive but (a) is not. Our first theorem in tournaments is due to Redei (1934).

THEOREM 5.4.1

Every tournament G has a directed Hamiltonian path.

Proof:

This is an immediate consequence of Theorem 5.3.4(c). However, an independent proof using induction on the number n of vertices is along the

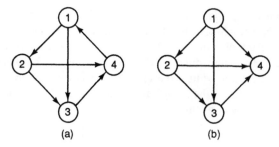

(a)   (b)   **FIGURE 5.4.1**

following lines. The theorem is true when $n = 2$. Suppose that it is true for n. Now consider any tournament G' with $(n + 1)$ vertices. Let v be any arbitrary vertex of G'. Now consider the subgraph G of G' obtained from G' by deleting v and all arcs from v and to v. Obviously, G is a tournament with n vertices and so it has a directed Hamiltonian path $v_1 - - > v_2 . . . > v_i . . . > v_n$. If there is an arc in G' from v to v_1 or from v_n to v, then G' has a directed Hamiltonian path and we are done. Otherwise let i be the largest integer such that there is no arc from v to v_i. So there is an arc from v_i to v. Now our choice of i is such that there is no arc from v_{i+1} to v implying that there is an arc in the opposite direction and consequently we have the directed Hamiltonian path in G' as follows: $v_1 . . . > v_i - - > v - - - > v_{i+1} . . . > v_n$, showing that the result is true for $(n + 1)$. $\diamondsuit$

COROLLARY

A transitive tournament has a unique directed Hamiltonian path.

Proof:

If P and P' are two distinct directed Hamiltonian paths, there is a pair of vertices x and y such that there is a path from x to y in P and a path from y to x in P'. So, by transitivity, there is an arc from x to y in P and an arc from y to x in P'. This is a contradiction. So $P = P'$. $\diamondsuit$

We conclude our discussion of tournaments with the following theorem on the existence of unique directed Hamiltonian paths in tournaments. See Roberts (1976) for a proof.

THEOREM 5.4.2

In a tournament G the following properties are equivalent:
(a) G has a unique directed Hamiltonian path.
(b) G has no directed cycles of length 3.
(c) G is acyclic.
(d) G is transitive.

5.5 VERTEX COLORING AND PLANARITY OF GRAPHS

A graph is said to be **colored** if each vertex is assigned a color such that no two adjacent vertices have the same color. If such an assignment of colors is possible using at most k colors, the graph is **k-colorable.** The smallest value of k such that a graph G is k-colorable is the **chromatic number** of G.

The chromatic number of a graph is 1 if and only if it has no edges. The chromatic number of a complete graph with n vertices is of course n, and the chromatic number of a bipartite graph is 2. In particular, the chromatic number of a tree is 2. Any cycle with p vertices is 2-colorable if and only if p is even. Consequently, if a graph G has an odd cycle (i.e., a cycle with an odd number of vertices), G is not 2-colorable. On the other hand, if there is no odd cycle in a graph G, the graph is 2-colorable. This is obvious if G is a tree because a tree is acyclic. More generally, assume that G is a connected graph with no odd cycles. Start from any vertex v and apply a breadth first search (BFS) procedure to obtain a BFS tree as follows: v is at level 0. All vertices adjacent to v are at level 1. We partition the vertices of the graph into sets of vertices at various levels. Let $v_{i1}, v_{i2}, \ldots, v_{ir}$ be the vertices at level i. Consider all vertices adjacent to v_{i1} that are not in levels $0, 1, 2, \ldots, i$. Put these vertices in a new level $(i + 1)$. Then consider all vertices that are adjacent to v_{i2} but not in levels $0, 1, 2, \ldots, i, (i + 1)$. Include these vertices in level $(i + 1)$. Continue this process until all the vertices are examined. We now assign two colors to the vertices: one color to all the vertices at the odd level and another color to all the vertices at the even level. In Figure 5.5.1 we have a graph with no odd cycles and a BFS tree starting from vertex 3 is as in Figure 5.5.2.

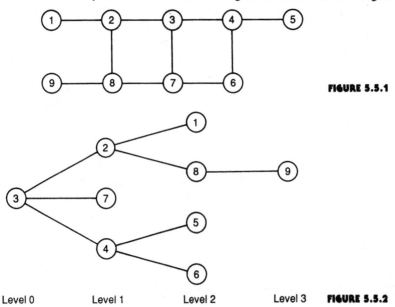

FIGURE 5.5.1

Level 0   Level 1   Level 2   Level 3   **FIGURE 5.5.2**

In the BFS procedure in the worst case we have to examine all the n vertices and all the m edges. So the worst-case complexity is $n + m$. Since m is at most $n(n - 1)/2$, the complexity is $n(n + 1)/2$.

From an earlier exercise we know that a graph is bipartite if and only if it has no odd cycles. Thus we have the following theorem to characterize the 2-colorability of graphs.

THEOREM 5.5.1

In a connected graph G, the following are equivalent:
(a) G is bipartite.
(b) G is 2-colorable.
(c) G has no odd cycles.

There is no known characterization for the k-colorability of graphs when $k > 2$. In general, it is a hard problem to compute the chromatic number of an arbitrary graph. There is no algorithm that always gives a coloring pattern using the fewest possible colors. However, there are algorithms to color a given graph that ''approximate'' the best coloring in the sense that it may sometimes use more colors than are absolutely necessary. Here is an algorithm, known as the **largest first algorithm,** because it assigns colors to vertice with the largest degrees first. First we order the vertices according to nonincreasing degrees. Use the first color to color the first vertex and then color, in sequential order, each vertex that is not adjacent to a previously colored vertex of the same color. Repeat this process using the second color for the subsequence of uncolored vertices. Continue this process until all vertices are colored.

Let us make use of the largest first algorithm to obtain a coloring for the graph in Figure 5.5.3. The nine vertices are labeled as $i = 1, 2, 3, \ldots, 9$, where the degree of i is greater than or equal to the degree of $(i + 1)$. First assign color 1 to vertex 1. Now vertices 8 and 9 are not adjacent for 1. So assign color 1 to vertex 8. We see that vertex 9 is not adjacent to vertex 8. So assign color 1 to vertex 9 as well. Thus vertices 1, 8, 9 are assigned color 1. The remaining vertices are 2, 3, 4, 5, 6, 7. Assign color 2 to vertex 2. The *uncolored* vertices not adjacent to vertex 2 are 3, 4, 6. Assign color 2 to vertex 3. We

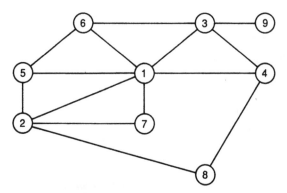

FIGURE 5.5.3

notice that vertices 3 and 4 are adjacent. So vertex 4 remains uncolored. Similarly, vertex 6 remains uncolored. Thus vertices 2 and 3 are colored with color 2.

The remaining uncolored vertices are 4, 5, 6, 7. Assign color 3 to vertex 4 and then to vertex 5 and vertex 7. Finally, we are left with vertex 6, to which assign color 4. Thus the graph of Figure 5.5.3 does not need more than four colors. But three colors will do the job. Assign color red to 1, color blue to 3, 5, 7, 8, and color green to 2, 4, 6, 9.

The idea of coloring a graph arises naturally in many scheduling problems. Suppose that the computer science department in a university has decided to offer a certain number of graduate courses and there are a certain number of time periods during which these courses can be offered. In scheduling these courses the department has to avoid conflicts. If a graduate student is interested in taking two of these courses, these courses have to be scheduled at different times. We construct a graph model of this scheduling problem as follows: Let each vertex represent a course. Join two vertices by an edge if the courses corresponding to these two vertices cannot be offered at the same time. If the resulting graph is k-colorable, the courses can be scheduled with k or more periods.

Planarity of Graphs

A graph is called a **planar graph** if it can be drawn so that no two edges intersect except at a vertex. A planar graph drawn on a plane so that no two edges intersect is a **plane graph.** The two-dimensional regions defined by the edges in a plane graph are the **faces** and the vertices and the various edges define the **boundaries** of these faces. In the plane graph of Figure 5.5.4 there are five vertices, seven edges, and four faces. In this plane graph, we see that F_1, F_2, and F_3 are the **interior faces,** and the unbounded region F_4 is the **exterior face.** The boundary of F_1 is the cycle 1- - - - -2- - - - -3- - - - -1 and the boundary of F_4 is 1- - - - -2- - - - -5- - - - -2- - - - -4- - - - -3- - - - -1, in which the last edge is e.

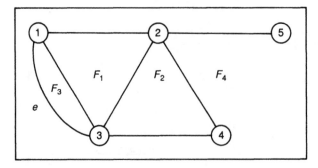

FIGURE 5.5.4

The classical (proved in 1750) result connecting the number of vertices, edges, and faces is the following theorem of Euler.

THEOREM 5.5.2

If a connected plane graph has n vertices and m edges, then the number of faces is p, where $n - m + p = 2$.

Proof:

We use induction on m. When $m = 0$, we have $n = 1$ and $p = 1$ and the result is true. Suppose that the result is true when $m = k - 1$. Consider any plane graph G with n vertices, k edges and p faces. We wish to show $n - k + p = 2$. This is certainly true if the graph is a tree. If it is not a tree, let e be any edge of a cycle. If we delete e, we still have a connected plane graph with n vertices, $(k - 1)$ edges, and $(p - 1)$ faces. Notice that if we delete an edge from a cycle two faces coalesce into one. By our induction hypothesis, $n - (k - 1) + (p - 1) = 2$, so $n - k + p = 2$, as we wished to prove.   ◇

We now establish another useful result, which is an immediate consequence of what we already proved.

THEOREM 5.5.3

A simple connected planar graph with n vertices (n is at least 3) has at most $(3n - 6)$ edges.

Proof:

If n is 3, the number of edges is at most three. Let n be greater than or equal to 3. We draw the plane graph with faces $F_1, F_2, \ldots, F_p$. Let r_i be the number of edges that define the face F_i. Then r_i is at least three for each i. So $3p \leq (r_1 + r_2 + \cdots + r_p)$. Now in counting the total number of edges in the boundaries, each edge is counted at most twice. Thus the right side of the inequality above is at most $2m$, where m is the number of edges in the graph. Hence $3p$ is at most $2m$. But by Theorem 5.5.2 we know that $p = 2 - n + m$. The result follows.   ◇

We use this theorem to establish the nonplanarity of some famous graphs. If a simple graph with n vertices has more than $(3n - 6)$ edges, it is nonplanar. The complete graph K_5 has five vertices and 10 edges, so it is not planar. We can establish the nonplanarity of the complete bipartite graph $K_{3,3}$ by contradiction. For this graph we know that $n = 6$ and $m = 9$. So if it is planar, it should have exactly five faces according to Theorem 5.5.2. Recall that a bipartite graph has no odd cycles. So there should be at least 20 edges to define the boundaries of these faces. Each edge is counted at most twice. So the graph should have at least 10 edges. But there are only 9. *Thus any graph that has K_5 or $K_{3,3}$ as a subgraph is nonplanar.* It turns out that every nonplanar graph contains one of these two as a subgraph in a certain sense which we now make precise.

Two graphs are said to be **homeomorphic** (or identical to each other within vertices of degree 2) if they both can be obtained from the same graph G by introducing new vertices of degree 2 on its edges. For example, the two graphs (a) and (b) in Figure 5.5.5 are homeomorphic. Notice that insertion or deletion of vertices on edges does not affect considerations of planarity. We now state the following celebrated theorem of Kuratowski (proved in 1930), which gives a necessary and sufficient condition for a graph to be planar. See Bondy and Murty (1976) for a proof.

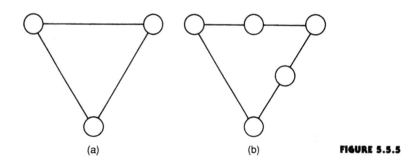

(a)   (b)   **FIGURE 5.5.5**

THEOREM 5.5.4

A graph is planar if and only if it contains no subgraph that is homeomorphic to K_5 or $K_{3,3}$.

Finally, a note on map coloring. When we color the different nations in a geographical map, two countries with a common border cannot have the same color. The map coloring problem then is to color a given map with as few colors as possible. No one has ever found a map that needs more than four colors. For more than 100 years it was conjectured that no map needs more than four colors, but no correct proof was forthcoming. Finally, in 1976, this four-color conjecture was settled. For a discussion, see Appel and Haken (1976).

Now given a geographical map we can construct a planar graph as follows. Consider each country as a vertex. Join two vertices by an edge if the two countries corresponding to these two vertices have a common border. The minimum number of colors required to color the map is the chromatic number of the graph thus constructed. Every map gives rise to a planar graph, and vice versa. Thus the four-color theorem can be reformulated as the following important theorem in graph theory.

THEOREM 5.5.5

The chromatic number of a planar graph cannot exceed four.

5.6 *NOTES AND REFERENCES*

Any book on graph theory will be a good reference for Eulerian and Hamiltonian graphs. The books by Behzad et al. (1979), Bondy and Murty (1976), Chartrand (1977), Deo (1974), Gibbons (1985), Gondran and Minoux (1984), Harary (1969a), Ore (1963), Roberts (1976, 1978), and Wilson (1979) are some of the standard ones. For the discussion of graph algorithms some excellent references are the books by Aho et al. (1983), Baase (1978), Even (1979), Gondran and Minoux (1984), Lawler (1976), and Reingold et al. (1977). The books by Minieka (1978), Papadimitriou and Steiglitz (1982), and Syslo et al. (1983) contain elaborate discussion of some well-known graph algorithms. Applications of graph theory to coding theory, operations research, computer science, and chemistry are presented in Deo (1974). For additional details on feedback shift registers (mentioned in Section 5.2), refer to the books by Golomb (1967) and Ronse (1982). The survey article by Ralston (1982) on de Bruijn sequences while demonstrating the connection between coding theory and graph theory also shows how different areas of discrete mathematics impinge on computer science. The survey paper by Bellmore and Nemhauser (1968) on the Traveling Salesman Problem is a good introductory reading, and the book on the same topic by Lawler et al. (1985) is a complete and systematic study of this celebrated topic. For a more general discussion on tournaments, see the book by Moon (1968).

Some excellent references on vertex coloring in graphs are the relevant chapters in the books by Behzad et al. (1979), Berge (1962), Chartrand (1977), Grimaldi (1985), Gould (1988), Harary (1969a), Liu (1985), Roberts (1976, 1978, 1984), and Wilson (1979). Comparable coverage of the material presented in this chapter is contained in Chapter 3 (Sections 3 and 4) and Chapter 6 (Section 1) of Roberts (1984). See the papers by Birkhoff and Lewis (1960) and Read (1968) for additional reading on chromatic numbers. The famous four-color conjecture originated in 1852 and was solved 124 years later when Appel and Haken (1976) demonstrated that every planar map can be colored with four or fewer colors. See the paper by Haken (1977) for a description of the proof of this theorem. Since their proof depended on dividing the problem into several cases depending on the arrangement of the countries in the map and analyzing the various colorings of these arrangements by writing computer programs, there is a controversy over the nature of this proof. See the paper by Tymoczko (1980) for some philosophical underpinnings regarding this controversy. Is there a purely mathematical proof without using any computer analysis showing that every map can be colored with four or fewer colors? This is still an open problem. An interesting historical account of this celebrated conjecture prior to its proof can be found in May (1965) and Harary (1969b).

5.7 EXERCISES

5.1. Find an Eulerian circuit in the graph of Figure 5.7.1.

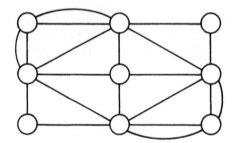

FIGURE 5.7.1

5.2. Prove that a graph is Eulerian if and only if its set of edges can be partitioned into cycles.

5.3. Show that a weakly connected digraph with an Eulerian circuit is strongly connected.

5.4. Show that a weakly connected digraph with an Eulerian path is unilaterally connected.

5.5. Can there be a bridge in an Eulerian graph?

5.6. Can there be a bridge in a graph which has an Eulerian path?

5.7. Construct a word with four letters A, B, C, and D using the following matrix:

$$
\begin{array}{c} \\ A \\ B \\ C \\ D \end{array}
\begin{array}{cccc} A & B & C & D \\ \left[\begin{array}{cccc} 0 & 1 & 0 & 1 \\ 1 & 0 & 0 & 0 \\ 0 & 1 & 0 & 1 \\ 1 & 0 & 1 & 1 \end{array}\right] \end{array}
$$

5.8. Construct a word with four letters A, B, C, and D with the frequency set {2, 1, 2, 4} using the following matrix:

$$
\begin{array}{c} \\ A \\ B \\ C \\ D \end{array}
\begin{array}{cccc} A & B & C & D \\ \left[\begin{array}{cccc} 1 & 0 & 0 & 1 \\ 0 & 0 & 0 & 1 \\ 0 & 0 & 1 & 1 \\ 1 & 1 & 1 & 0 \end{array}\right] \end{array}
$$

5.9. Draw the de Bruijn digraph $G(p, n)$ and obtain a de Bruijn sequence $B(p, n)$ when (a) $p = 3$, $n = 2$, and (b) $p = 3$, $n = 3$.

5.10. Draw a connected graph with five vertices that is Eulerian but not Hamiltonian.

5.11. Draw a connected graph with four vertices that is Hamiltonian but not Eulerian.

5.12. Draw a connected graph with four vertices that is both Eulerian and Hamiltonian.

5.13. Draw a connected graph with four vertices that is neither Eulerian nor Hamiltonian.

5.14. Prove Theorem 5.3.2.

5.15. Prove that a bipartite graph with an odd number of vertices is non-Hamiltonian.

5.16. If there is a Hamiltonian path from vertex i to vertex j in a digraph G, then i is a "winner" and j is a "loser" in G. Construct a tournament with five players such that (a) each player can be a winner as well as a loser, (b) there is a unique winner and a unique loser.

5.17. Is every tournament unilaterally connected? What can you say about the converse? Justify your answers.

5.18. Use the largest first algorithm to color the vertices in the graph of Figure 5.7.2.

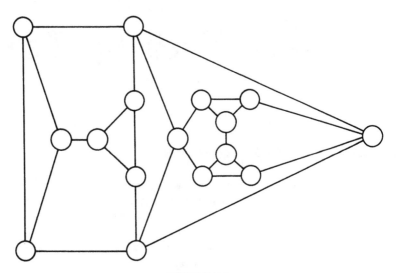

FIGURE 5.7.2

CHAPTER

6

Trees
and Their Applications

6.1 DEFINITIONS AND PROPERTIES

A connected graph with no cycles is a **tree.** One of the more widely studied discrete structures is that of a tree. Trees are especially suited to represent hierarchical structures and addresses and labels. Special types of trees are used in coding theory and in searching. We will be seeing some of these applications in this chapter. Before doing this we shall establish a few results pertaining to the characterization of trees.

1. *In a tree there is a unique simple path between every pair of vertices.* We prove this assertion as follows. Let u and v be two vertices in a tree T. Since T is connected there is a path between u and v and therefore there is a simple path between them. If possible, let P and P' be two simple paths between them. If the two paths are not the same, there is an edge in P that is not in P'. Let us assume that e is the first edge that we come across when we go from u to v that is in P but not in P'.

Let

$$P: \quad u. \quad . . . . .u_i\text{-} \text{-}\overset{e}{\text{-}}\text{-} \text{-}u_{i+1}. \quad . . . . .v$$

and

$$P': \quad u. \quad . . . . .u_i\text{-} \text{-} \text{-} \text{-}v_{i+1}. \quad . . . . .v$$

Let W be the set of intermediate vertices in P between u_{i+1} and v, and let W' be the set of intermediate vertices in P' between v_{i+1} and v. If W and W' have no elements in common, then we get a cycle starting from u_i going through all the vertices in W, vertex v, and then all the vertices in W'. On the other hand, if W and W' have a common vertex, let r be the least subscript of a vertex u_r in P such that u_r is in W'. So none of the intermediate vertices in P between u_i and u_r is in P'. Then we have a cycle starting from u_i which goes through all the vertices in W up to u_r and then all the vertices in W' from u_r to u_i. Thus the existence of two distinct simple paths between two vertices implies the existence of a cycle. By definition, a tree is acyclic. So there is a unique path between every pair of vertices in a tree.

2. *Conversely, if there is a unique simple path between every pair of vertices in a graph G, then G is a tree.* Suppose that G is not a tree. Then there is at least one cycle C in G which implies that between any two vertices in C there are two simple paths and this is a contradiction.

3. In a tree T, an edge between two vertices v and w is the unique path between them, and if we delete this edge from T, then T is no more connected. *In other words, every edge in a tree is a bridge.*

4. *Conversely, if G is a connected graph such that every edge is a bridge, then G is a tree.* Suppose that G is not a tree and let C be a cycle in G. Let e be any edge in C. Let G' be the subgraph of G after deleting e. Since e is a bridge G' is no more connected. Let p and q be any two vertices in G. There is a path P between p and q in G. If P does not contain e, then P is a path between p and q in the (disconnected) graph G' as well. On the other hand, if $e = (v, w)$ is an edge in P that is also in the cycle C which starts from the vertex t, we have the following path in G' between p and q:

$$p. \quad . . . . .v. \quad . . . . .t. \quad . . . . . .w. \quad . . . . .q$$

In other words, there is a path between every pair of vertices in G', and this contradicts the fact that G' is not connected.

5. *A tree T with n vertices has $(n-1)$ edges.* We prove this by induction on n. This is true when $n = 1$. Suppose that it is true for all m, where $1 < m < n$. Let $e = \{u, w\}$ be an edge in T. Since T is a tree, e is a bridge. Delete e to obtain the subgraph T' which has two connected components, H and H'. Both H and H' are trees with k and k' vertices. Now k and k' are positive

integers whose sum is n. So they both are less than n. By our induction hypothesis H has $(k - 1)$ edges and H' has $(k' - 1)$ edges, and together they have $k + k' - 2 = n - 2$ edges. So T' has $(n - 2)$ edges, and consequently, T has $(n - 1)$ edges.

6. The converse of (5) is true: *any connected graph G with n vertices and $(n - 1)$ edges is a tree.* For if $G = (V, E)$ is not a tree, there is an edge e that is not a bridge. We delete e to get a connected subgraph $G' = (V, E')$. Continue thus until we get a subgraph $H = (V, F)$ in which every edge is a bridge. So H is a tree with $(n - 1)$ edges. Let $k (> 0)$ be the number of edges removed from G in this process. We see that after deleting k edges from $(n - 1)$ edges we are left with $(n - 1)$ edges!

7. Our next assertion is *that any acyclic graph $G = (V, E)$ with n vertices and $(n - 1)$ edges is connected and therefore a tree.* Suppose that G is not connected. Let the components of G be G_i (with n_i vertices) for $i = 1, 2, \ldots, r$. Now each component G_i is acyclic and connected and therefore a tree with $n_i - 1$ edges. Thus the total number of edges in $G = n_1 + n_2 + \cdots + n_r - r$ that is equal to $n - r$. But G has exactly $n - 1$ edges. So $r = 1$. That is, G has exactly one acyclic component.

8. *Let G be any graph with n vertices. If any two of the following statements are true, then the third is also true*: (a) G is connected, (b) G is acyclic, and (c) G has $(n - 1)$ edges.

9. Let T be any tree. Join any two nonadjacent vertices v and w by a new edge, resulting in a graph G. Then G has exactly one cycle, consisting of the new edge and the unique simple path in T between v and w. On the other hand, if G is an acyclic graph such that whenever any two nonadjacent vertices are joined by a new edge the resulting graph has exactly one cycle, then G is a tree. The proof is by contradiction. Suppose that G is not a tree. Then it is not connected. So there is a pair of vertices p and q in G such that there is no path between them and so the addition of the new edge $\{p, q\}$ does not create a cycle.

10. Let G be any connected graph such that whenever two nonadjacent vertices are joined by a new edge, the resulting graph has exactly one cycle. Then G is acyclic and therefore a tree.

We now summarize this relentless sequence of assertions as three theorems.

THEOREM 6.1.1

The following are equivalent in a simple graph G:

(a) G is connected and acyclic.

(b) G is *connected* and the number of edges in G is one less than the number of vertices in it.

(c) G is *acyclic* and the number of edges in G is one less than the number of vertices in it.

(d) G is connected and every edge is a bridge.

(e) There is a unique simple path between every pair of vertices in G.

(f) G is *acyclic* and if any two nonadjacent vertices are joined to construct G', then G' has exactly one cycle.

(g) G is *connected* and if any two nonadjacent vertices are joined to construct G', then G' has exactly one cycle.

DEFINITION 6.1.1

A subgraph T of a graph G with n vertices is a **spanning tree** in the graph if (a) T is a tree and (b) T has n vertices.

THEOREM 6.1.2

A graph G is connected if and only if it has a spanning tree.

THEOREM 6.1.3

Let G be a simple graph with n vertices. If a subgraph H with n vertices satisfies any two of the following three properties, then it satisfies the third as well.

(a) H is connected.

(b) H has $(n - 1)$ edges.

(c) H is acyclic.

(Notice that Theorem 6.1.3 characterizes spanning trees in a graph. In Chapter 4 we used the depth-first search procedure to obtain a spanning tree in a connected simple graph.)

6.2 SPANNING TREES

In Chapter 4 we mentioned that graph theory was "born" in 1736. In the same vein one could say that trees were first used by G. B. Kirchhoff (1824–1877) in 1847 in his work on electrical networks. Analysis of an electrical network actually reduces to finding all spanning trees of the graph under consideration. Spanning trees also form the basis for a large number of problems in network optimization. Some of these problems are taken up in Chapter 7. Even though Kirchhoff used trees in his analysis, it was Arthur Cayley (1821–1895) who, a decade later, used trees systematically in his attempts to enumerate the isomers of the saturated hydrocarbons (compounds of the form C_kH_{2k+2}), which can be represented as a connected graph with $3k + 2$ vertices (one for each carbon atom C and one for each hydrogen atom H). Since the valences of C and H are

4 and 1, the sum of all the degrees is $4k + (2k + 2)$ and therefore there should be $3k + 1$ edges. Thus the graph under consideration is, in fact, a tree. In other words, the graph T of a hydrocarbon with k carbon atoms is a spanning tree in an arbitrary graph with $3k + 2$ vertices such that the degree of each C vertex is 4 and the degree of each H vertex is 1. The natural question to ask: How many distinct hydrocarbons can exist for a given value of k? In this context Cayley proved a theorem on the number of spanning trees in a graph. A tree with n vertices is called a **labeled tree** if each vertex is assigned a unique label i where i is a positive integer between 1 and n. Two labeled trees are *distinct* if their edge sets are different. For example, 1- - -2- - -3, 1- - -3- - -2, and 2- - - 1- - -3 are three distinct labeled trees when $n = 3$, whereas 1- - -2- - -3 and 3- - -2- - -1 are not distinct labeled trees.

THEOREM 6.2.1

The number of distinct labeled trees (with n vertices) is n^{n-2} (where n is at least 2).

Proof:

Let N' be the set of all $(n - 2)$-tuples of $N = \{1, 2, \ldots, n\}$. Each element in N' has $n - 2$ components and each component can be chosen in n ways. So the cardinality of N' is n^{n-2}. Our theorem is proved if we establish a one-to-one correspondence between N' and the set of distinct labeled trees with n vertices. Let T be any labeled tree with n vertices, and let W be the set of vertices in T of degree 1. (A vertex of degree 1 in a tree is a **leaf**.) W has at least two and at most $n - 1$ elements. Arrange the elements of W in an increasing order and let w_1 be the first element in W. Let s_1 be the unique vertex adjacent to w_1. Next, let T' be tree obtained by deleting w_1 from T, and let W' be the set of vertices of degree 1 in T' arranged in an increasing order. If w_2 is the first element in W', we take s_2 as the unique vertex adjacent to w_2 in T'. We continue this process until we get a $(n - 2)$-tuple s of the form $(s_1 \quad s_2 \quad s_3 \quad \cdots \quad s_{n-2})$ establishing that every labeled tree corresponds to a unique element in N'.

Before we establish the result in the opposite direction, let us actually obtain a 10-tuple for a labeled tree T with 12 vertices as shown in Figure 6.2.1. $W = \{5, 6, 7, 8, 9, 10, 11, 12\}$ and $s_1 = 1$, which is the vertex adjacent to the first element in W. Delete from T the vertex 5 and the edge joining 1 and 5 to obtain the tree T' in which W' is the set of all vertices of degree 1 arranged in an increasing order. The first element in W' is 1, so s_2 is 4.

Next, delete vertex 1 and the edge joining 1 and 4. The first vertex of degree 1 in the new tree is 6, which is adjacent to 2. Thus s_3 is 2. We continue similarly and observe that $s_4 = 2$, $s_5 = 4$, $s_6 = 3$, $s_7 = 3$, $s_8 = 3$, $s_9 = 4$,

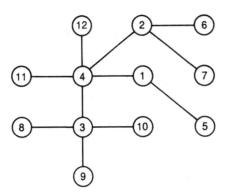

FIGURE 6.2.1

and finally, $s_{10} = 4$. Thus the labeled tree T corresponds to the 10-tuple
(1 4 2 2 4 3 3 3 4 4).

Next, we prove that every $(n - 2)$-tuple s defines a unique labeled tree
with n vertices. If

$$s = (s_1 \quad s_2 \quad s_3 \quad \cdots \quad s_{n-2})$$

we define

$\quad v_1 =$ the first element in N that is not in s
$\quad v_2 =$ the first element in $N - \{v_1\}$ that is not in $s - \{s_1\}$
$\quad v_3 =$ the first element in $N - \{v_1, v_2\}$ that is not in $s - \{s_1, s_2\}$, etc.

We repeat this process until we get v_i ($i = 1, 2, 3, \ldots, n - 2$). The
two remaining elements in N are denoted by x and y. Now construct a graph
whose vertex set is N for which the edges are the $(n - 2)$ edges joining s_i and
v_i and the edge joining x and y. The graph thus obtained is the unique labeled
tree that corresponds to s.

For example, if

$$s = (1 \quad 4 \quad 2 \quad 2 \quad 4 \quad 3 \quad 3 \quad 3 \quad 4 \quad 4)$$

then $n = 12$ and $N = \{1, 2, 3, 4, 5, 6, 7, 8, 9, 10, 11, 12\}$. Thus $v_1 =$ first
element in N that is not in s. So $v_1 = 5$. Next $v_2 =$ first element in $N - \{5\}$
that is not in $s - \{1\}$. So $v_2 = 1$. $v_3 =$ first element in $N - \{5, 1\}$ that is not
in $s - \{1, 4\}$. So $v_3 = 6$. Continuing like this, we get $v_4 = 7$, $v_5 = 2$, $v_6 = 8$, $v_7 = 9$, $v_8 = 10$, $v_9 = 3$, and $v_{10} = 11$. Finally, $x = 4$ and $y = 12$. Now
construct the graph with vertex set N and edges joining s_i and v_i ($i = 1, 2, 3, \ldots, 10$) and the edge joining x and y. This graph is precisely the labeled tree

in Figure 6.2.1. Thus the one-to-one correspondence between N' and the set of distinct labeled trees with n vertices is established proving the theorem.   ◇

(*Note*: Cayley's theorem is an existence theorem. It does not solve the problem of finding all the distinct labeled trees.)

6.3 BINARY TREES

A digraph is called a **directed tree** if its underlying graph is a tree. A directed tree is a **rooted tree** if (1) there is exactly one vertex (called the **root**) with indegree 0 and (2) the indegree of every other vertex is 1. A vertex in a rooted tree is a **terminal vertex** (or a **leaf**) if its outdegree is 0. A nonterminal vertex is called an **intermediate** or **internal vertex.** The root is thus an intermediate vertex. The number of arcs in the path from the root to a vertex is called the **level** of that vertex. By definition the level of the root is 0. If the levels are 0, $1, 2, \ldots, k$, then k is the **height** of the tree. Vertex v is a **descendant** of vertex u if there is an arc from u to v. If T is a rooted tree, a **subtree** T' **at vertex** v is a rooted tree $T' = (V', E')$ such that (1) the root of T' is at v, (2) V' consists of v and all its descendants in T, and (3) E' contains all the arcs of all the directed paths (in T) from v to all the leaves. A tree is **pruned at vertex** v if we delete all the descendants of v and all the arcs of all the directed paths emanating from v so that v becomes a leaf of the pruned tree.

A rooted tree is a **binary tree** if the outdegree of each intermediate vertex is at most 2. In a **regular binary tree** the outdegree of each intermediate vertex is exactly 2. A regular binary tree is **full** if all its leaves are at the same level. Many discrete structures can be represented as binary trees. We discuss here an application in coding theory.

In a computer, since all information is stored in the form of binary numbers, whenever a letter in the alphabet or a symbol is entered into the computer it is converted into a binary word by means of a character code that is a one-to-one correspondence between the set of all characters and a set of binary numbers. For example, two of the most commonly used character codes are the ASCII (American Standard Code for Information Interchange) and EBCDIC (Extended Binary-Coded Decimal Interchange Code). In both these codes, each character is assigned a code of fixed length and therefore they are known as **fixed-length character codes.** In ASCII the length is usually 8, and it is 6 in EBCDIC. Thus to encode the number 247 we use a message of 24 bits: 011100100111010001110111. Decoding is also straightforward. The number of bits in a message is a multiple of 8 and we use the code systematically to decipher the message.

The chief drawback of a fixed-length character code is that all characters, whether they are used frequently or not, need the same number of bits. It is certainly advantageous to have a code in which more frequently used characters use fewer bits so that the total length of the message is as small as possible. In other words, a **variable-length character code** is more appropriate. But while using such a code, we have to make sure that the decoding is unambiguous. For example, if the codes for 1, 2, and 3 are 01, 0, and 00, respectively, then the word 0001 can be decoded as 221 or 31. An obvious reason for this ambiguity is the fact that the code for 2 appears as a part of the code for another number in its beginning. A word w is a **prefix** of another word v if $v = wp$, where p is another word. A character code is said to have the **prefix property** if no code for a symbol is a prefix of the code for another symbol. A character code is a **prefix code** if it has the prefix property.

It is easy to see that a binary prefix code can be obtained straight from a regular binary tree. First we label the two arcs incident from an intermediate vertex as 0 and 1. Then assign to each leaf a sequence of bits that is the sequence of labels of the arcs from the root to that leaf. A character code that assigns each character into the sequence which corresponds to a leaf is necessarily a prefix code. For example, the assignment $A = 00$, $B = 010$, $C = 011$, $D = 100$, $E = 101$, and $F = 11$ obtained from the regular binary tree of Figure 6.3.1 is a prefix code for the alphabet {A, B, C, D, E, F}.

Conversely, corresponding to a given prefix code we can construct a regular binary tree by pruning a full regular binary tree of length k, where k is the length of the largest sequence in the code. For example, consider the code R = 00, E = 01, A = 10, and D = 111 with the prefix property. Here $k = 3$ and first we construct a full regular binary tree of height 3 and label each arc emanating from any intermediate vertex as 0 or 1. Then assign to each vertex of the tree a sequence of bits that is the sequence of labels of the arcs from the root to that vertex. Thus each binary sequence of length less than or equal to k is assigned to a unique vertex. Locate those vertices whose binary sequences are precisely equal to the sequences in the given prefix code and prune the tree at these vertices.

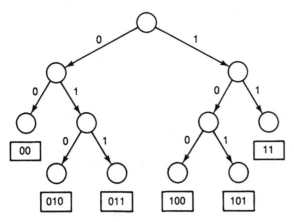

FIGURE 6.3.1

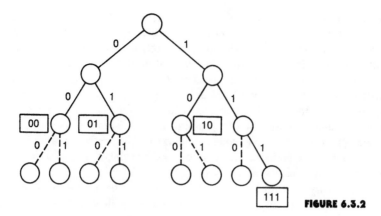

FIGURE 6.3.2

Also delete leaves that do not correspond to sequences in the code. The binary tree of Figure 6.3.2 is obtained for the given prefix code after locating the appropriate vertices.

Thus given a prefix code an arbitrary string can be unambiguously decoded by proceeding from left to right in the string, finding the first substring that is a character, then the subsequent substring, and so on. For example, if the prefix codes are 00, 010, 011, 100, 101, and 11, the unique deciphering of the string 10111100100000100111110 is 101, 11, 100, 100, 00, 010, 011, 11. The last two bits are not used in this case.

Finally, for a given alphabet it is possible to have more than one prefix code. For example, both {00, 01, 10, 11} and {0, 11, 100, 101} are prefix codes for {A, B, C, D}. So it is natural to ask whether one code is "better" than the other. It is in this context that we consider the *efficiency* of a code.

Let $\{a^1, a^2, \ldots, a^n\}$ be an alphabet and let f^i be the frequency with which a^i appears on an average. Let C be any prefix code for the alphabet and let l^i be the length of the code for a^i. The weight $w(C)$ of the code C is $f^1 l^1 + \cdots + f^n l^n$. The problem thus is to find a code C such that $w(C)$ is as small as possible and it is equivalent to the following problem: Given that f^i ($i = 1, 2, \ldots, n$) is the frequency of the character a^i in an alphabet of n letters, find a regular binary tree with n terminal vertices such that $f^1 l^1 + f^2 l^2 + \cdots + f^n l^n$ is a minimum where l^i is the length of the path from the root to the ith terminal vertex, which represents a_i. Such a tree is known as an optimal binary tree for the given frequency distribution. We now discuss an elegant procedure due to D. Huffman (1952) to obtain such a tree.

First we arrange the frequencies in nondecreasing order from f^1 to f^n. Huffman's algorithm is based on the following fact (see Theorem 6.3.1 below): If T' is an optimal tree obtained using this procedure for the $(k - 1)$ frequencies $\{f^1 + f^2, f^3, \ldots, f^k\}$ with $(k - 1)$ terminal vertices, the tree T obtained from T' by introducing two new terminal vertices v_1 (to represent f^1) and v_2 (to represent

f^2) and by joining the terminal vertex v in T' which represents $f^1 + f^2$ to v_1 and v_2 is an optimal tree for the set $\{f^1, f^2, f^3, \ldots, f^k\}$. (The sum of f^1 and f^2 need not be less than f^3. Tiebreaking is arbitrary in the sense that if $f^1 + f^2 = f^3$, then v can be either the vertex that represents $f^1 + f^2$ or the vertex that represents f^3. Notice that v is not a terminal vertex in T.)

Here is an example that illustrates the construction of a Huffman code for an alphabet of six letters with a frequency distribution $S^1 = \{3, 5, 6, 8, 10, 14\}$. Add the first two numbers and rearrange to get $S^2 = \{6, 8, 8, 10, 14\}$. Repeat this process to get $S^3 = \{8, 10, 14, 14\}$, $S^4 = \{14, 14, 18\}$, and $S^5 = \{18, 28\}$. Now construct an optimal tree for S^5 as in Figure 6.3.3. Now the frequency 28 is the sum of 14 and 14 in S^4, and this will take us to S^4 and the tree in Figure 6.3.4. Then we use the decomposition $18 = 8 + 10$ to get a tree for S^3 as in Figure 6.3.5. After that we use the decompositions $14 = 6 + 8$ (once) and $8 = 3 + 5$ to get the desired Huffman tree, as in Figure 6.3.6.

Thus a Huffman code for the frequency distribution $\{3, 5, 6, 8, 10, 14\}$ is $\{000, 001, 110, 111, 01, 10\}$ with lengths 3, 3, 3, 3, 2, 2, and the weight of this code is $3.3 + 3.5 + 3.6 + 3.8 + 2.10 + 2.14 = 114$, which is a minimum.

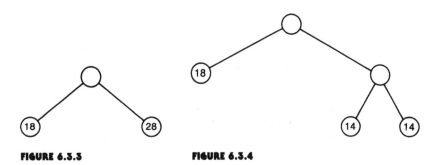

FIGURE 6.3.3   **FIGURE 6.3.4**

THEOREM 6.3.1

The regular binary tree obtained by using the Huffman algorithm is an optimal tree.

Proof:

Let the n frequencies f^i ($i = 1, 2, \ldots, n$) be in a nondecreasing order. The number of regular binary trees with n terminal vertices representing these n frequencies is finite, and therefore there is a regular binary tree T' with n vertices for which the weight $w(T')$ is a minimum. Let v be an internal vertex of T' such that the distance (i.e., number of edges in the path) from the root of the tree to v is not less than the distance from the root to any other nonterminal

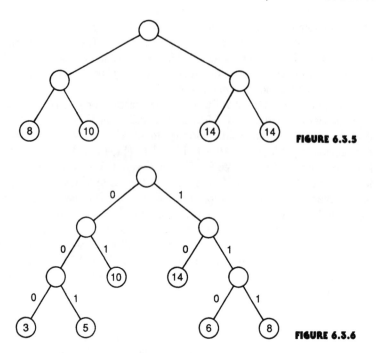

FIGURE 6.3.5

FIGURE 6.3.6

vertex, and let the descendants of v be the terminal vertices representing f^i and f^j. Now assign the value f^1 to the vertex that represents f^i and the value f^2 to the vertex that represents f^j. At the same time, assign the value f^i to the vertex that represents f^1 and the value f^j to the vertex that represents f^2. The weight $w(T)$ of the resulting tree T cannot be more than $w(T')$ because of the choice of v. At the same time $w(T)$ cannot be less than $w(T')$ because T' is an optimal tree. Hence we conclude that there is an optimal tree T for which f^1 and f^2 are the descendants of the same nonterminal vertex that is at a maximal (in comparison with other nonterminal vertices) distance from the root of T.

Next, let T'' be the regular binary tree with $(n - 1)$ terminal vertices obtained from the tree T obtained above by deleting the two terminal vertices that correspond to f^1 and f^2, so that the vertex v (from which these two vertices descend) becomes a terminal vertex, which corresponds to the frequency $f^1 + f^2$. It is easy to see that $w(T) = w(T'') + f^1 + f^2$. Thus T is optimal if and only if T'' is optimal. The conclusion of the theorem follows by induction on n. ◇

We conclude this discussion by establishing that Huffman's algorithm indeed gives a coding system that has the desired property: The larger the frequency of a character, the shorter the length of the code that represents that character.

THEOREM 6.3.2

Let T be an optimal Huffman tree for the n characters a^i with frequencies f^i, and let d^i be the length of the code that represents a^i ($i = 1, 2, \ldots, n$). If $f^i < f^j$, then $d^i \geq d^j$.

Proof:

The length from the root of T to the terminal vertex that represents a^i is d^i. Let the terminal vertex that represents f^i be given the code a^j, and let the terminal vertex that represents f^j be given the code a^i. Even though the tree is the same, its weight will change. Let w be the old weight and w' be the new weight. Of course, $w \leq w'$. Now $w - w' = (d^i - d^j)(f^i - f^j)$. Thus $d^i \geq d^j$.   ◇

Binary Search Trees

An important operation that frequently arises in computer science is the use of binary trees to search through a list or a table, the elements of which constitute a linearly ordered finite set. If T is a binary tree and if the outdegree of a vertex v is 2, then v has two descendants—a **left descendant** and a **right descendant.** (If the out degree of v is 1, the unique descendant may be considered as either the left descendant or the right descendant.) The tree rooted at the left descendant of a vertex is called the **left subtree of this vertex.** The tree rooted at the right descendant of a vertex is called the **right subtree of this vertex.**

Suppose that T is a binary tree. To each vertex v of the tree, a real number $k(v)$ is assigned. This number $k(v)$ is called the **key** of the vertex. Assign keys to the vertices of T such that the key of a vertex is (1) larger than the keys of the vertices in its left subtree and (2) smaller than the keys of the vertices in its right subtree. A binary tree with keys defined by this rule is called a **binary search tree.** In Figure 6.3.7 we have a binary search tree with nine vertices in which the set of keys is $\{9, 7, 13, 4, 8, 11, 15, 3, 5\}$.

If T is an arbitrary binary tree, it is possible to assign keys to the vertices of T such that T is a binary search tree. This is proved in Reingold (1977), in which an algorithm to convert a binary tree into a binary search tree is also derived. To search whether a particular item q is in a list, we use the binary search tree of the list as follows: First we start with the root. At any stage, the key of the vertex x is examined. In the beginning, x is the root. If $q = k(x)$, we found the desired item in the list. If $q < k(x)$, we ignore the right subtree rooted at x and look at the left descendant of x. If $q > k(x)$, we ignore the left subtree rooted at x and look at the right descendant of x. We continue this process until we reach a tree with one vertex.

In this search process, the number of comparisons to be made will be

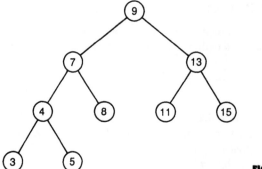

FIGURE 6.3.7

$h + 1$ in the worst case where h is the height of the binary search tree. Thus the computational complexity of this algorithm is minimized if we can find a binary search tree of minimum height. In this context, the following theorem is very handy.

THEOREM 6.3.3

The minimum height of a binary tree with n vertices is $m - 1$ where m is the ceiling of $\log(n + 1)$.

Proof:

Let T be any binary tree with n vertices, and let h be its height. There are at most 2^k vertices at level k, where $k = 0, 1, 2, \ldots, h$. So

$$n \leq 1 + 2 + 2^2 + 2^3 + \cdots + 2^h = 2^{h+1} - 1$$

Hence $h + 1 \geq \log(n + 1)$, which implies that $h \geq m - 1$, where m is the ceiling of $\log(n + 1)$.   ◇

A recursive procedure can be used to *construct a binary search tree* corresponding to a given linearly ordered list as follows. Start with a tree containing just one vertex, which is the root. We assign the first file in the list as the key of this vertex. To add a new file from the list, this file is first compared with the keys of the existing vertices in the tree, starting at the root and moving to the left if the file is less than the key of the vertex under consideration and if this vertex has a left descendant, or moving to the right if the file is greater than the key of the vertex under consideration and if this vertex has a right descendant. When the current file is less than the key of a vertex and if this vertex has no left descendant, a new vertex with this file as its key is inserted as a new left

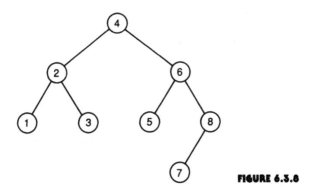

FIGURE 6.3.8

descendant. Similarly, when a file is greater than the key of a vertex and if this vertex has no right descendant, a new vertex is added with this file as its key as a new right descendant. This procedure is illustrated in the following example.

Example 6.3.1

Form a binary search degree to represent {4, 6, 2, 8, 5, 3, 7, 1}.

Solution   The binary search tree is displayed in Figure 6.3.8. The key of the root is 4, which is the first element in the set. The next element, 6, is more than 4. Since 4 has no right descendant at this stage, a new vertex with key 6 is created as the right descendant of 4. The next element is 2. We start with the root and move left since 2 < 4. Since 4 has no left descendant, a new vertex is constructed as the left descendant of 4, and this vertex has the key 2. The next element is 8. Compare 8 with 4 and we move right. Compare 8 with 6 and we move right. Now the vertex with key as 6 has no right descendant. Now a new vertex with key 8 is constructed as the right descendant of the vertex with key 6. We continue this process until we reach the last item in the list.

6.4 *NOTES AND REFERENCES*

Some useful general references on trees and their properties are the appropriate sections from the books by Behzad et al. (1979), Berge (1962), Bondy and Murty (1976), Deo (1974), Gondran and Minoux (1984), Gould (1988), Grimaldi (1985), Harary (1969a), Liu (1985), Roberts (1984), Townsend (1987), Tucker (1984), and Wilson (1979). Theorem 8.2.1 is due to Arthur Cayley (1821–1895), who used graph theory in connection with enumeration problems in physical chemistry. For additional treatment of Huffman codes, see Huffman (1952),

Markowsky (1981), and Standish (1980). See Chapter 2 of Knuth (1973a) and Chapter 6 of Knuth (1973b) for a complete treatment of trees and search trees.

6.5 EXERCISES

6.1. If G is a forest with n vertices, m edges, and k components, obtain an expression for m in terms of n and k.

6.2. Suppose that a tree has two vertices of degree 5, three vertices of degree 4, six vertices of degree 3, eight vertices of degree 2, and r vertices of degree 1. Find r.

6.3. G is a connected graph with 20 vertices. Find the minimum number of edges that G can have.

6.4. G is a connected graph with 20 edges. Find the maximum number of vertices that G can have.

6.5. Suppose G has four components, 20 edges, and r vertices, Find the maximum value of r.

6.6. Show that every tree is a bipartite graph. Which trees are complete bipartite graphs?

6.7. An edge e in G is in every spanning tree of G. What can you say about e?

6.8. An edge e in G is in no spanning tree of G. What can you say about e?

6.9. If T and T' are two spanning trees in G, is it necessary that T and T' have an edge in common? Either prove this or produce a counterexample.

6.10. Show that a connected graph in which every vertex is even must have a cycle.

6.11. If e is an edge in a connected graph G (with no loops), then prove that there is a spanning tree $T(e)$ which contains e.

6.12. If e and f are two edges in a simple graph, there is a spanning tree $T(e, f)$ that contains both e and f.

6.13. Find the unique labeled tree that corresponds to $s = (8 \quad 8 \quad 7 \quad 7 \quad 7 \quad 6 \quad 6)$.

6.14. If the edges of a labeled tree are $\{1, 2\}$, $\{2, 3\}$, $\{2, 4\}$, $\{4, 5\}$, $\{4, 7\}$, $\{5, 6\}$, $\{7, 8\}$, and $\{8, 9\}$, find s.

6.15. Use the prefix code $A = 000$, $B = 001$, $C = 01$, $D = 10$, $E = 111$, and $R = 110$ to decode the following word:
000001110000010001000000111000011

6.16. If the frequencies of the six letters in the prefix code in Problem 6.15 are 8, 10, 4, 5, 12, and 10, find the weight of the code.

6.17. Obtain an optimal prefix code for the data in Problem 6.16 and find the weight of this code. Encode the word that appears in Problem 6.15 using this code. What is the length of this word if we use this optimal code in the worst case?

6.18. Decode the word 11111011001001 using the optimal code obtained in Problem 6.17.

6.19. Find a regular binary tree of height k with 13 vertices such that **(a)** k is as small as possible, and **(b)** k is as large as possible.

6.20. A rooted tree of height k is said to be a **balanced tree** if every terminal vertex is at level k or $(k - 1)$. Construct a balanced binary tree with n vertices when $n = 11$ and $n = 12$.

6.21. Construct a binary search tree (using alphabetical order) for the set {Hungary, Germany, Poland, Bulgaria, Romania, Czechoslovakia, Albania, Yugoslavia}.

CHAPTER

7

Spanning Tree Problems

7.1 MORE ON SPANNING TREES

If G is a connected graph with n vertices, a spanning tree in G, as we saw in Chapter 6, is an acyclic subgraph of G with $(n - 1)$ edges. If $T = (V, E')$ is a spanning tree in $G = (V, E)$, the edges of G not in E' are called the **chords** of T. If e is a chord joining the vertices u and v, the edges in the *unique path in T between u and v* together with the edge e form a *unique* cycle in G which is called the **fundamental cycle of G relative to T with respect to the chord** e and is denoted by $C^T(e)$. Thus if G has m edges, it will have $m - (n - 1)$ such fundamental cycles relative to every spanning tree. We assume that G is a connected graph throughout this chapter unless otherwise stated.

A subset D of the set of edges of a graph $G = (V, E)$ is called a **disconnecting set** of G if the deletion of the edges in D from G makes G into a disconnected graph. If V is partitioned into two sets V' and V'' and if $D = (V', V'')$ is the set of all edges in E of the form $\{i, j\}$, where $i \in V'$ and $j \in V''$, then D is a disconnecting set.

A disconnecting set D is called a **cutset of** G if no proper subset of D is a disconnecting set. A disconnecting set consisting of exactly one edge is, of course, a cutset known as a **bridge.**

If D is a cutset in G, the deletion of the edges in D disconnects G into *exactly two* components G' (with V' as the set of vertices) and G'' (with V'' as the set of vertices), and thus $D = (V', V'')$. On the other hand, if V is arbitrarily partitioned into two sets W and W', the disconnecting set $D = (W, W')$ need not be a cutset. For example, let $V = \{a, b, c\}$, $E = \{\{a, b\}, \{a, c\}\}$, $W = \{a\}$, and $W' = \{b, c\}$. In this case the disconnecting set $D = (W, W')$ is not a cutset. So when will a partition of the vertex set give rise to a cutset?

THEOREM 7.1.1

If the vertex set V of a connected graph G is partitioned into two subsets W and W' such that every two vertices in W (and W') are connected by a path that consists of vertices only from W (and W'), then $D = (W, W')$ is a cutset.

Proof:

Suppose that D is not a cutset. Then there is proper subset D' of D that is a cutset. Let $e = \{w, w'\}$ be an edge in D that is not in D', where w is in W and w' is in W'. Suppose that u and v are any two vertices in G, where u is in W and v is in W'. By hypothesis there is a path between u and w consisting of vertices from W only, and there is a path between w' and v consisting of vertices from W' only. Thus there is a path between u and v using the edge e that is not in D'. Thus D' is not a disconnecting set, which is a contradiction. $\Diamond$

COROLLARY

If T is any spanning tree in $G = (V, E)$, the deletion of any edge in T makes T disconnected by creating two subtrees with vertex sets W and W' such that $D(W, W')$ is a cutset of G.

Thus corresponding to each edge e of a spanning tree T, there is a unique cutset $D^T(e)$ called the **fundamental cutset** of T with respect to the edge e. Thus any connected graph with n vertices will have a system of $(n - 1)$ fundamental cutsets with respect to every spanning tree. For example, in the graph of Figure 7.1.1 we have a spanning tree T with edges a, b, c, d, and e. The chords are p, q, r, and s. T thus has four fundamental cycles (one with respect to each chord of T) and five fundamental cutsets (one with respect to each edge of T). We see that the edges in the fundamental cycle $C(r)$ are r, c, d, and e. The edges in the cutset $D(b)$ are b, p, and q.

There is a close relation between the concepts of spanning trees, cycles, and cutsets, and this is the content of the next two theorems.

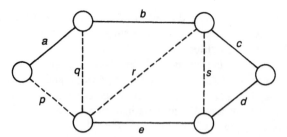

FIGURE 7.1.1

THEOREM 7.1.2

Let T be a spanning tree in a connected graph G, and let C and D be a cycle and cutset, respectively, in G. Then:

(a) Either there are no edges in common between C and D or there are an even number of edges in common between the two.

(b) At least one edge of C is a chord of T.

(c) At least one edge of D is an edge of T.

Proof:

(a) Let $D = (W, W')$. If all the vertices of C are in one of the two subsets, then of course C and D have no edges in common. Suppose that w and w' are two vertices in C where w is in W and w' is in W'. Then the cycle C that starts from w and ends in w will necessarily use the edges from D an even number of times: Whenever an edge from D of the form $\{i, j\}$, where i is in W and j is in W', is used, an edge of the form $\{u, v\}$ is also used, where u is in W' and v is in W.

(b) If no edge of C is a chord of T, then C *is a subgraph of* T, which is a contradiction since T is acyclic.

(c) If no edge of D is an edge of T, the deletion of the edges from D will not disconnect G because such a deletion will not affect the spanning tree. ◇

THEOREM 7.1.3

(a) Let $D(e)$ be the fundamental cutset with respect to an edge e of a spanning tree T, and let f be any other element in this cutset. Then (1) f is a chord of T defining the fundamental cycle $C(f)$, (2) e is an element of $C(f)$, and (3) if e' is another chord of T that is not in $D(e)$, then e is not an element of $C(e')$.

(b) Let $C(e)$ be the fundamental cycle with respect to the chord e of a spanning tree T, and let f be any other edge in this cycle. Then (1) f is an edge of the tree defining the fundamental cutset $D(f)$, (2) e is an element of $D(f)$, and (3) if e' is another edge of T that is not in $C(e)$, then e is not an element of $D(e')$.

Proof of (a):

(1) Any edge f in $D(e)$, other than e, cannot be an edge in T since T is acyclic. So f is a chord defining a fundamental cycle $C(f)$.

(2) Let f be any edge in $D(e)$ other than f. Then $D(e) = \{e, f\} \cup A$, where A is a set of chords of T, and $C(f) = \{f\} \cup B$, where B is a set of edges of T. The edge f is common for both $D(e)$ and $C(f)$. Recall that $D(e)$ and $C(f)$ should have an even number of edges in common. Since A and B have no edges in common, we conclude that e is the only other edge common to both $D(e)$ and $C(f)$. Thus e is an edge in the fundamental cycle $C(f)$.

(3) Let $C(e')$ be the fundamental cycle with respect to e' that is a chord not in $D(e)$. Thus $C(e') = e' \cup L$, where L is a set of edges of T, and $D(e) = e \cup M$, where M is a set of chords of T. If e is in $C(e')$, then e' will be in $D(e)$, which is against our assumption. So e is not in $C(e')$.

Proof of (b):

This proof is similar to that of (a) and is left as an exercise.

We can restate parts (a) and (b) of Theorem 7.1.3 as follows:

(a) If e is any edge of a spanning tree T in a connected simple graph G, then (1) there is a unique cutset $D(e)$; (2) if f is any edge in this cutset other than e, then f is a chord of T that defines a unique cycle $C(f)$ such that e is an edge in this cycle; and (3) if e' is a chord of T that is not in $D(e)$ defining a unique cycle $C(e')$ then e is not an edge in $C(e')$.

(b) If e is any chord of a spanning tree T in a connected simple graph G, then (1) there is a unique cycle $C(e)$; (2) if f is any edge in this cycle other than e, then f is an edge of T that defines a unique cutset $D(f)$ such that e is an edge in this cutset; and (3) if e' is an edge of T that is not in $C(e)$ defining a unique cutset $D(e')$, then e is not an edge in $D(e')$.

If we associate a real number (called the **weight**) with each edge of a graph G so that G becomes a network, the **weight of a spanning tree** T in G is then the sum of the weights of all the edges in T. A spanning tree T is a **minimal spanning tree** (MST) if the weight of T does not exceed the weight of any other spanning tree in G. The MST problem (also known as the minimal connector problem) is the problem of finding a MST in a connected graph G. This optimization problem has several important practical applications. For example, it can be helpful in planning large-scale communication and distribution

networks when the most important consideration usually is to provide paths
between every pair of vertices in the most economical way. The vertices would
be cities, terminals, or retail outlets, and the edges would be highways or pipe-
lines. The weights corresponding to these edges could be distances or costs or
time involved in these processes. See Graham and Hell (1982) for an exhaustive
survey and many references. In this chapter we present two simple algorithms
to find a minimal spanning tree in a graph. One is due to Kruskal (1956), and
the other is due to Prim (1957). The approach in both the methods is *greedy*: It
so happens that if we take the "choicest morsel" at each opportunity without
violating any rules, we will have eventually an optimal solution.

The minimal spanning tree problem has the following generalization. Let
W be a subset of the set V of all vertices of a connected simple graph G. A tree
$T = (U, F)$ in G, where W is a subset of U, is called a **Steiner tree with respect
to the set** W. The **minimal Steiner network problem for the set** W is the
problem of finding a Steiner tree with respect to W of minimum weight. Thus
a minimal Steiner tree with respect to V is a minimal spanning tree in G. It is
quite possible that a minimal Steiner tree with respect to a proper subset W is a
minimal spanning tree in G. There is no known efficient algorithm to solve the
Steiner tree problem. An efficient algorithm to obtain an approximate solution
is presented in Chang (1972).

7.2 KRUSKAL'S GREEDY ALGORITHM

We list the edges of the connected network with n vertices in an ascending order
of *nondecreasing weights* and then construct a subgraph T examining these edges
one at a time starting with an edge of the smallest weight. An edge will be added
to T as long as it does not form a cycle with some or all the edges of T. The
construction halts when T has $(n - 1)$ edges. Obviously, this greedy procedure
ensures that T is a spanning tree. That the T thus obtained is indeed a *minimum*
spanning tree is a consequence of the following result.

THEOREM 7.2.1

If e is an edge in a cycle C of a connected graph G such that the weight
of e is more than the weight of any other edge in the cycle C, then e is not an
edge for any MST in G.

Proof:

Suppose that T is a MST in which e is an edge. Let $D(e)$ be the fundamental
cutset with respect to e. Since the edge e is common for both the cycle C and
the cutset $D(e)$, there should be at least one more element f common to both

these sets because the number of elements common to a cycle and a cutset is even. Since f is in $D(e)$, f is necessarily a chord of T. Let $C(f)$ be the fundamental cycle with respect to f. By Theorem 7.1.2 we know that e is an element of $C(f)$. Now consider the subgraph H obtained by adjoining f to T. The only cycle in T is $C(f)$, and if we delete e from H, we get a spanning tree T' with weight less than that of T. This contradiction establishes the fact that e is not an edge of T.

$\diamondsuit$

COROLLARY

If the weight of any other edge in C does not exceed the weight of e, there is a minimal spanning tree in which e is not an edge.

In Kruskal's algorithm we abandon an edge p in favor of another edge q for inclusion in T (when the weight of p does not exceed that of q) only when the inclusion of p creates a cycle in which p is an edge with the largest weight. Thus Kruskal's algorithm correctly solves the MST problem. It is also an easy exercise at this stage to verify that if all the weights of the edges in G are distinct, there is a unique MST in G.

As an example, consider the graph of Figure 7.2.1, where we have the sorted list L of all the edges of G as

$$L = \{\{1, 2\}, (1, 5\} \{2, 5\}, \{2, 3\}, \{3, 5\}, \{3, 6\}, \{5, 6\}, \{1, 4\}, \{4, 5\}\}$$

in an ascending order. The algorithm examines $\{1, 2\}$ and accepts it for the tree. Then it examines $\{1, 5\}$ and accepts it. After that it examines $\{2, 5\}$ and does not accept it, to avoid the cycle C consisting of $\{1, 2\}$, $\{1, 5\}$ and $\{2, 5\}$. Then it proceeds further and accepts $\{2, 3\}$, $\{3, 6\}$, and $\{1, 4\}$ in turn. At this stage it halts because the number of edges accepted is one less than the total number of vertices. In this graph we have nine edges and we had to examine eight of them before we stop.

How many computational steps (in this case, comparisons) are needed to arrange (to sort) the m edges of the graph with nondecreasing order? The number of such comparisons no doubt depends on the algorithm we use to sort the m elements of the edge set E. One obvious method is as follows: Successively compare the ith term to the $(i + 1)$st term in the set, interchanging the two if the ith term is larger than the $(i + 1)$st term. This procedure is called the **bubblesort** because the larger numbers "rise" to the top. The first number in the set has to be compared with at most $(m - 1)$ numbers. Then the second number has to be compared with at most $(m - 2)$ numbers, and so on. Thus the total number of comparisons in the worst case if we use the bubblesort algorithm is $1 + 2 + 3 + \cdots + (m - 1) = m(m - 1)/2$, which is a polynomial in m of degree 2. In other words the worst-case complexity of the

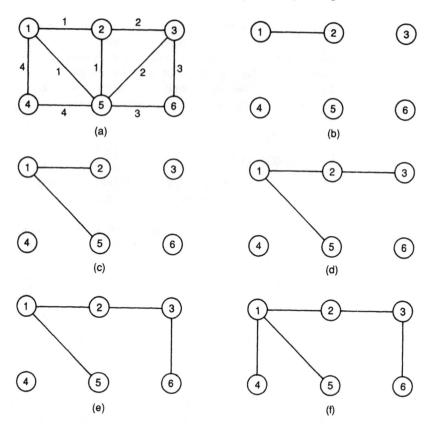

FIGURE 7.2.1

bubblesort algorithm is $O(m^2)$. See the Appendix for notations and concepts related to computational complexity of algorithms. On the other hand if we use the **mergesort** algorithm (see Stanat and McAllister, 1977) the number of computations to sort m numbers in the worst case is $O(m \log m)$. In general, there is no sorting algorithm that is more efficient than this. Another well-known algorithm known as **heapsort** has a worst-case behavior of $O(m \log m)$, whereas the **quicksort** algorithm has an average-case behavior of $O(m \log m)$ and a worst-case behavior of $O(n^2)$, where n is the number of vertices. See Aho et al. (1983) for more details. Under these circumstances it is reasonable to conclude that the worst-case complexity of Kruskal's algorithm is $O(m \log m)$. Notice, however, that if m is very large in comparison with n [i.e., when m is $O(n^2)$], it is not very economical to sort all the m edges when we need only $(n - 1)$ of these m edges. See Syslo et al. (1983) for implementation details of Kruskal's algorithm when m is large.

7.3 PRIM'S GREEDY ALGORITHM

In this procedure we start with an arbitrary vertex v in G and examine all the edges incident at v. Let $e = \{v, w\}$ be an edge with least weight among all these edges. We construct a subgraph T of G starting with e as an edge. Next examine all edges (other than e) incident at v and all edges incident at w and choose an edge f of least weight among them. This newly found edge is added to the subgraph T. The edge f is either between v and a new vertex or between w and a new vertex. Let u be the new vertex. At this stage we have three vertices v, w, and u. Examine all the edges other than e and f that are incident at v, u, and w and choose the one with the smallest weight such that the edges e and f and the newly selected edge g do not form a cycle. At this stage g is added to T. We continue until all vertices are accounted for. This procedure obviously ends up with a spanning tree. That this tree is indeed an MST is a consequence of the two corollaries of the following theorem.

THEOREM 7.3.1

If v is any vertex in a connected network G and if e is an edge incident at v such that the weight of e is less than the weight of every edge incident at v, then e is an edge of every minimum spanning tree in G.

Proof:

Let T be a MST and suppose that $e = \{v, w\}$ is not an edge of T. Let H be the subgraph of G obtained by adding e to T. H has a unique cycle $C(e)$ that can be represented as $v\text{- - - -}v_1\text{- - - -}v_2\text{- - - -}. \ \ . \ . \ . \ .v_r\text{- - - -}w\text{- - - - -}v$, where $e = \{v, w\}$ and let $f = \{v, v_1\}$. Now both e and f are incident at v and the weight of e is less than that of f. If we remove f from H we get a spanning tree T' with weight less than the weight of T, which is a contradiction.   ◇

COROLLARY 1

If v is any vertex of G and if e is an edge incident at v such that the weight of no edge incident at v is less than the weight of e, there is an MST in G for which e is an edge.

COROLLARY 2

If a tree T' that spans the vertices in a subset W of vertices in a connected graph $G = (V, E)$ is a subtree of a minimal spanning tree of G, there is a minimal spanning tree of G that contains T' and the smallest edge connecting W and $V - W$.

At each iteration of Prim's algorithm we have a partition of the vertex set $V = \{1, 2, \ldots, n\}$ into subsets P and Q where P is the set of vertices already accounted for and Q is its complement. Initially, we take $P = \{1\}$. We associate a label $t(i)$ for each vertex i in Q. Initially, $t(i)$ is the weight of the edge between 1 and i if there is an edge; otherwise, it is infinity (a large positive number). In step 1 we choose a vertex v in Q with the smallest label. Then we locate a vertex u in P such that the weight $d(u, v)$ of the edge between u and v is $t(v)$. At this point the edge $\{u, v\}$ is accepted as an edge for the MST and v is added to P. In step 2 we update the labels of the remaining vertices in Q as follows. If w is in Q, define $t(w) := \text{Min}\{t(w), d(v, w)\}$, where v is the latest entry in P. We continue similarly until $P = V$.

The worse-case complexity of the algorithm can be obtained immediately. Initially, Q has $(n - 1)$ elements. So in step 1, there will be at most $(n - 2)$ comparisons to start with. Thus this step entails $(n - 2) + (n - 3) + \cdots + 2 + 1$ comparisons. In step 2 we have $(n - 2)$ elements to start with. The label of each vertex w in Q has to be compared with $d(v, w)$, where v is the latest entry in P. This involves $(n - 2)$ comparisons to start with and then $(n - 3)$ comparisons. Thus step 2 also entails in the worst case as many comparisons as in step 1. Thus the worst-case complexity of Prim's algorithm is twice the sum of the first $(n - 2)$ natural numbers, which is $O(n^2)$. The different iterations of this algorithm for the network of Figure 7.3.1 are as follows.

Iteration 1

Step 1:
$P = \{1\}$ $Q = \{2, 3, 4, 5, 6\}$

$$
\begin{cases}
t(2) = 1 \\
t(3) = - \\
t(4) = 4 \\
t(5) = 1 \\
t(6) = -
\end{cases}
$$

A smallest label corresponds to vertex 2. So the edge $\{1, 2\}$ is in T. The set P is updated as $P = \{1, 2\}$.

Step 2:
$P = \{1, 2\}$ $Q = \{3, 4, 5, 6\}$

$$
\begin{cases}
t(3) = \text{Min}\{t(3), d(2, 3)\} = 2 \\
t(4) = \text{Min}\{t(4), d(2, 4)\} = 4 \\
t(5) = \text{Min}\{t(5), d(2, 5)\} = 1 \\
t(6) = \text{Min}\{t(6), d(2, 6)\} = -
\end{cases}
$$

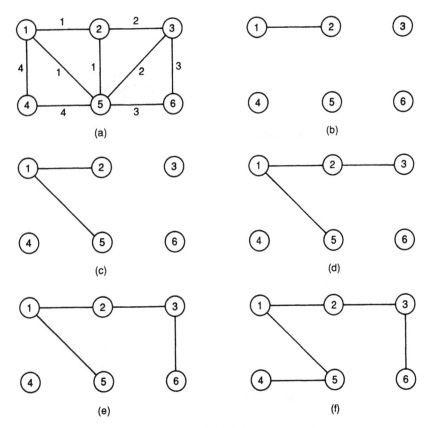

FIGURE 7.3.1

Iteration 2

Step 1:
$P = \{1, 2\}$ $Q = \{3, 4, 5, 6\}$

$$
\begin{cases}
t(3) = 2 \\
t(4) = 4 \\
t(5) = 1 \\
t(6) = -
\end{cases}
$$

Vertex 5 is chosen for updating P. Since $d(1, 5) = d(2, 5)$, we may take either the edge $\{1, 5\}$ or the edge $\{2, 5\}$ for updating T.

Step 2:
$P = \{1, 2, 5\}$ $Q = \{3, 4, 6\}$

$$\begin{cases} t(3) = \text{Min } \{t(3), d(5, 3)\} = 2 \\ t(4) = \text{Min } \{t(4), d(5, 4)\} = 4 \\ t(6) = \text{Min } \{t(6), d(5, 6)\} = 3 \end{cases}$$

Iteration 3

Step 1:
$P = \{1, 2, 5\}$ $Q = \{3, 4, 6\}$

$$\begin{cases} t(3) = 2 \\ t(4) = 4 \\ t(6) = 3 \end{cases}$$

Vertex 3 is chosen for updating P and the edge $\{2, 3\}$ is chosen for updating T.

Step 2:
$P = \{1, 2, 5, 3\}$ $Q = \{4, 6\}$

$$\begin{cases} t(4) = \text{Min } \{t(4), d(3, 4)\} = 4 \\ t(6) = \text{Min } \{t(6), d(3, 6)\} = 3 \end{cases}$$

Iteration 4

Step 1:
$P = \{1, 2, 3, 5\}$ $Q = \{4, 6)\}$

$$\begin{cases} t(4) = 4 \\ t(6) = 3 \end{cases}$$

Vertex 6 is chosen for updating P and the edge $\{3, 6\}$ is chosen for updating T.

Step 2:
$P = \{1, 2, 3, 5, 6\}$ $Q = \{4\}$

$$t(4) = \text{Min } \{t(4), d(6, 4)\} = 4$$

Iteration 5

Step 1:
Vertex 4 is chosen for updating P and the edge $\{5, 4\}$ is chosen for updating T.

Step 2:
$P = \{1, 2, 3, 4, 5, 6\}$ Q = the empty set.

Output: The edges of an M.S.T. are $\{1, 2\}$, $\{1, 5\}$, $\{2, 3\}$, $\{3, 6\}$, and $\{4, 5\}$.

Prim's Algorithm (Matrix Method)

Let $D = (d(i, j))$ be the $n \times n$ matrix, where n is the number of vertices of G and $d(i, j)$ is the weight of the edge $\{i, j\}$ if there is an edge between i and j. Otherwise, it is infinity. Initially we delete all elements of column 1 and mark row 1 with a *. All elements initially are with no underlines. Each iteration has two steps as follows.

Step 1: Select a smallest element from the entries (with no underlines) in the starred rows. Stop if no such element exists. The edges that correspond to the underlined entries constitute a MST.

Step 2: If $d(i, j)$ is selected in step 1, underline that entry, mark row j with a *, and delete the remaining elements in column j. Go to step 1.

As an illustration, let us consider the network as given in Figure 7.2.1. Initially, we have

$$D = \begin{bmatrix} - & 1 & - & 4 & 1 & - \\ - & - & 2 & - & 1 & - \\ - & 2 & - & - & 2 & 3 \\ - & - & - & - & 4 & - \\ - & 1 & 2 & 4 & - & 3 \\ - & - & 3 & - & 3 & - \end{bmatrix} \begin{matrix} * \\ \\ \\ \\ \\ \end{matrix}$$

in which all entries in column 1 have been deleted and row 1 is starred. At this point no entry is underlined.

Iteration 1:

$$D = \begin{bmatrix} - & \underline{1} & - & 4 & 1 & - \\ - & - & 2 & - & 1 & - \\ - & - & - & - & 2 & 3 \\ - & - & - & - & 4 & - \\ - & - & 2 & 4 & - & 3 \\ - & - & 3 & - & 3 & - \end{bmatrix} \begin{matrix} * \\ * \\ \\ \\ \\ \end{matrix}$$

Iteration 2:

$$
D = \begin{bmatrix}
- & \underline{1} & - & 4 & \underline{1} & - \\
- & - & 2 & - & - & - \\
- & - & - & - & - & 3 \\
- & - & - & - & - & - \\
- & - & 2 & 4 & - & 3 \\
- & - & 3 & - & - & -
\end{bmatrix}
\begin{matrix} * \\ * \\ \\ \\ * \\ \end{matrix}
$$

Iteration 3:

$$
D = \begin{bmatrix}
- & \underline{1} & - & 4 & \underline{1} & - \\
- & - & \underline{2} & - & - & - \\
- & - & - & - & - & 3 \\
- & - & - & - & - & - \\
- & - & - & 4 & - & 3 \\
- & - & - & - & - & -
\end{bmatrix}
\begin{matrix} * \\ * \\ * \\ \\ * \\ \end{matrix}
$$

Iteration 4:

$$
D = \begin{bmatrix}
- & \underline{1} & - & 4 & \underline{1} & - \\
- & - & \underline{2} & - & - & - \\
- & - & - & - & - & \underline{3} \\
- & - & - & - & - & - \\
- & - & - & 4 & - & - \\
- & - & - & - & - & -
\end{bmatrix}
\begin{matrix} * \\ * \\ * \\ \\ * \\ * \end{matrix}
$$

Iteration 5:

$$
D = \begin{bmatrix}
- & \underline{1} & - & \underline{4} & \underline{1} & - \\
- & - & \underline{2} & - & - & - \\
- & - & - & - & - & \underline{3} \\
- & - & - & - & - & - \\
- & - & - & - & - & - \\
- & - & - & - & - & -
\end{bmatrix}
\begin{matrix} * \\ * \\ * \\ \\ * \\ \end{matrix}
$$

The procedure at this stage halts, giving the edges that correspond to the underlined entries in the matrix D of the last iteration. Thus the edges $\{1, 2\}$, $\{1, 4\}$, $\{1, 5\}$, $\{2, 3\}$, and $\{3, 6\}$ form a minimal spanning tree in G.

7.4 COMPARISON OF THE TWO ALGORITHMS

The execution time of Prim's algorithm depends only on the number of vertices, but the time for Kruskal's algorithm increases as the number of edges is increased for a network with the same number of vertices. However, in general, it is not possible to assert which one is more efficient. The efficiency depends on, among other things, the structure of the network and the distribution of weights. Many variations, based primarily on data structures and implementation details, have been suggested to improve the efficiency. It has been observed that for networks with vertices up to 100, Prim's method appears to be more efficient, particularly so when there is an abundance of edges. The following running times for the two algorithms run on an AMDHL 470 V/8 computer are reported by Syslo et al. (1983).

Number of vertices	Number of edges	Execution time (msec)	
		Kruskal	Prim
80	790	64	52
80	1580	95	51
80	3160	186	50
100	200	44	78
100	300	66	80

7.5 NOTES AND REFERENCES

For a discussion of spanning trees in general, see any standard book on graph theory listed at the end of the book. See Lawler (1976), where an algorithm ("not a very good one") is described to solve the Steiner tree problem. For the MST problem the earliest references are probably the classical papers of Kruskal (1956) and Prim (1957). The matrix description of Prim's algorithm is from Hu (1982). For a description of the implementation details of these two greedy algorithms, a very good reference is the book by Syslo et al. (1983). Other general references are Cheriton and Tarjan (1976), Graham and Hell (1982), and Gabow et al. (1986).

7.6 EXERCISES

7.1. Suppose that G is a graph in which T is a spanning tree, C is a cycle, and D is a cutset. Prove the following:

(a) C and D have an even number of edges in common.

(b) D and T have at least one edge in common.

(c) C and the complement of T have at least one edge in common.

7.2. Prove that if the weights of the edges in a connected graph are all distinct, there is a unique MST in the graph.

7.3. If e is the unique edge in a connected network with the smallest weight, prove that e is an edge in every MST in G.

7.4. (a) Obtain a Steiner tree with respect to the vertex set $W = \{1, 2, 4, 5\}$ in the network shown in Figure 7.2.1.

(b) Suppose that the weights of the edges $\{1, 5\}$, $\{2, 5\}$, and $\{4, 5\}$ are all 10 units each. What will be the Steiner tree with respect to W?

7.5. Use Kruskal's algorithm to obtain a MST in G with the following weight matrix:

$$\begin{bmatrix} 0 & 20 & 15 & 4 & 3 & - \\ 20 & 0 & 19 & - & - & 9 \\ 15 & 19 & 0 & 8 & - & 10 \\ 4 & - & 8 & 0 & 6 & 9 \\ 3 & - & - & 6 & 0 & 7 \\ - & 9 & 10 & 9 & 7 & 0 \end{bmatrix}$$

7.6. Use Prim's algorithm (matrix method) to obtain a MST in the graph of Problem 7.5. Do you get the same tree in Problem 7.5 and in Problem 7.6? Find the weights of the two trees.

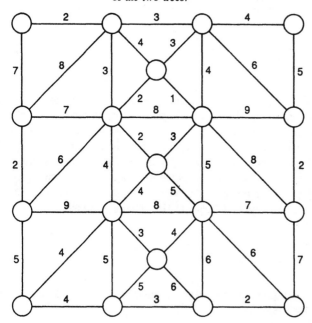

FIGURE 7.6.1

7.7. Modify Kruskal's algorithm to obtain a maximum spanning tree.

7.8. Obtain a maximum spanning tree in the graph of Problem 7.5.

7.9. Delete as many edges as possible from the graph G of Figure 7.6.1 to get a connected graph G' such that the weight of G' is minimum.

7.10. A cycle in a connected network G that passes through every vertex of G is called a Hamiltonian cycle. An arbitrary connected network G need not have a Hamiltonian cycle. The sum of the weights of all the edges in a Hamiltonian cycle is the weight of the Hamiltonian cycle, and a Hamiltonian cycle with minimum weight is called a traveling salesman (TS) cycle. Show that it is possible to obtain a lower bound for the weight of a TS cycle in a connected network G using the MST algorithm whether or not G has a Hamiltonian cycle. Obtain such a bound for the network in Problem 7.10.

7.11. Construct a network with 5 vertices and 5 edges of distinct weights in which the unique MST is the minimal Steiner tree with respect to 4 of these vertices.

CHAPTER

8

Shortest Path
Problems

8.1 INTRODUCTION

If each arc of a digraph is assigned a numerical weight (i.e., a distance), it is a natural and intuitively appealing problem to find a shortest path (if it exists) from a prescribed vertex to another prescribed vertex. Many optimization problems can be formulated and solved as shortest path problems of this type, and many complex problems in operations research can be solved by procedures that call upon shortest path algorithms as subroutines. Shortest path problems are in fact the most fundamental and also the most commonly encountered problems in combinatorial optimization. According to Goldman (1982), a shortest path algorithm developed by the U.S. Department of Transportation is regularly used billions of times every year. We confine our attention to two types of problems: (1) the problem of finding a shortest path from a vertex v to another vertex w, and (2) the problem of finding a shortest path from every vertex to every other vertex. Of course, (1) is a special case of (2).

In what follows we discuss two polynomial algorithms to solve the shortest path (S.P.) problem. The first algorithm is to find a S.P. and the shortest distance (S.D.) from a specified vertex to every other vertex. This algorithm is due to Dijkstra (1959). Our next algorithm, known as the Floyd–Warshall algorithm, enables us to find the S.P. and S.D. from every vertex to every other vertex. This procedure is due to Floyd (1964) and Warshall (1962). We assume that the weight function is nonnegative in the case of Dijkstra's algorithm even though it is possible to relax this restriction. It should be noted that there is a real difference between problems involving nonnegative weight functions and problems involving arbitrary weight functions. In the latter case the problem becomes unbounded if the network has a cycle with negative weight. The Floyd–Warshall algorithm detects the existence of such negative cycles.

8.2 DIJKSTRA'S ALGORITHM

In the network $G = (V, E)$, let $V = \{1, 2, \ldots, n\}$ and let the weight of the arc (i, j) be $a(i, j)$, which is assumed to be nonnegative. If there is no arc from i to j ($i \neq j$), then $a(i, j)$ is taken as infinity. We thus have the $n \times n$ weight matrix $A = (a(i, j))$ for G, in which all diagonal numbers are 0. The problem is to find the S.D. and S.P. from vertex 1 to all other vertices.

The procedure is as follows. Each vertex i is assigned a label that is either permanent or tentative. The permanent label $L(i)$ of i is the S.D. from 1 to i, whereas the tentative label $L'(i)$ of i is an upper bound of the S.D. from 1 to i. At each stage of the procedure, P is the set of vertices with permanent labels and T is its complement. Initially, $P = \{1\}$ with $L(1) = 0$ and $L'(i) = a(1, i)$ for each i. When $P = V$ the algorithm halts. Each iteration consists essentially of two steps, as follows.

Step 1 (Designation of a Permanent Label):
    Find a vertex k in T for which $L'(k)$ is minimal. Stop if there is no such k because then there is no path from 1 to any vertex in T. Adjoin k to the set P. Stop if $P = V$.

Step 2 (Revision of Tentative Labels):
    If j is a vertex in T, replace $L'(j)$ by the smaller value of $L'(j)$ and $L(k) + a(k, j)$. Go to step 1.

We now prove that the algorithm correctly solves the problem by induction on the number of elements in P.

THEOREM 8.2.1

Dijkstra's algorithm finds the S.D. from 1 to each i.

Proof:

We prove by induction on the number of elements in P that for every i in P, $L(i)$ is equal to the S.D. from 1 to i, and for every j not in P, $L'(j)$ is the length of a S.P. from 1 to j, every intermediate vertex of which is in P. This is true when P has one element. Suppose that this is true when P has up to m elements. By induction hypothesis just before vertex k is adjoined to the set P, $L'(k)$ is equal to the length of a S.P. from 1 to k in which every intermediate vertex is a vertex in P. Now k is adjoined to P and $L(k) = L'(k)$. We claim that $L(k)$ is the S.D. from 1 to k. If not, let d be the S.D. from 1 to k. Then $d < L(k) = L'(k)$. So any S.P. from 1 to k should have at least one vertex not from P as an intermediate vertex. Let v be the first such vertex. Let d' be the S.D. from 1 to v. Then, obviously, $d' \le d$. But $d < L'(k)$, which implies that $d < L'(k)$. This contradicts the assumption that $L'(k)$ is minimal.   ◇

The worse-case complexity of the algorithm is $O(n^2)$. This can be established as follows. At most there are n iterations. In Step 1, in the first iteration we have at most $(n - 2)$ comparisons, in the next iteration we have at most $(n - 3)$ comparisons, and so on. Therefore, there will be at most $(n - 2) + (n - 3) + \cdots + 1$ comparisons in Step 1. In Step 2 we again have at most $(n - 2)$ comparisons and also $(n - 2)$ additions in the first iteration. Thus in Step 2 we have $(n - 2) + (n - 3) + \cdots + 1$ comparisons and an equal number of additions in the worst case. Thus in all we have $(n - 1)(n - 2)$ comparisons and $(n - 1)(n - 2)/2$ number of additions, establishing the polynomial complexity of the algorithm.

Once the S.D. from 1 to i is known it is easy to find a S.P. from 1 to i by examining vertices j such that (1) $L(j)$ is less than $L(i)$ and (2) there is an arc from j to i. Here is an illustrative example to find the S.D. and S.P. from vertex 1 to the remaining vertices in a directed network as shown in Figure 8.2.1.

Iteration 1

Step 1:

$P = \{1\}$ $T = \{2, 3, 4, 5, 6, 7\}$

$L(1) = 0$
$$\begin{cases} L'(2) = 4 \\ L'(3) = 6 \\ L'(4) = 8 \\ L'(5) = - \\ L'(6) = - \\ L'(7) = - \end{cases}$$

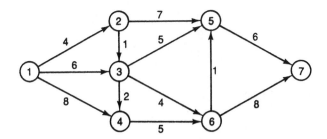

FIGURE 8.2.1

Vertex 2 gets a permanent label.

Step 2:
$P = \{1, 2\}$ $T = T - \{2\} = \{3, 4, 5, 6, 7\}$

$\begin{cases} L(1) = 0 \\ L(2) = 4 \end{cases}$

$\begin{cases} L'(3) = \text{Min}\,\{6,\ L'(2) + a(2, 3)\} \\ L'(4) = \text{Min}\,\{8,\ L'(2) + a(2, 4)\} \\ L'(5) = \text{Min}\,\{-, L'(2) + a(2, 5)\} \\ L'(6) = \text{Min}\,\{-, L'(2) + -\} \\ L'(7) = \text{Min}\,\{-, L'(2) + -\} \end{cases}$

Iteration 2

Step 1:
$P = \{1, 2\}$ $T = \{3, 4, 5, 6, 7\}$

$\begin{cases} L(1) = 0 \\ L(2) = 4 \end{cases}$

$\begin{cases} L'(3) = 5 \\ L'(4) = 8 \\ L'(5) = 11 \\ L'(6) = - \\ L'(6) = - \end{cases}$

Vertex 3 gets a permanent label.

Step 2:
$P = \{1, 2, 3\}$ $T = T - \{3\} = \{4, 5, 6, 7\}$

$\begin{cases} L(1) = 0 \\ L(2) = 4 \\ L(3) = 5 \end{cases}$

$\begin{cases} L'(4) = \text{Min}\,\{8, L'(3) + a(3, 4)\} \\ L'(5) = \text{Min}\,\{11, L'(3) + a(3, 5)\} \\ L'(6) = \text{Min}\,\{-, L'(3) + a(3, 6)\} \\ L'(7) = \text{Min}\,\{-, L'(3) + -\} \end{cases}$

Iteration 3

Step 1:
$P = \{1, 2, 3\}$ $T = \{4, 5, 6, 7\}$

$$\begin{cases} L(1) = 0 \\ L(2) = 4 \\ L(3) = 5 \end{cases}$$

$$\begin{cases} L'(4) = 7 \\ L'(5) = 10 \\ L'(6) = 9 \\ L'(7) = - \end{cases}$$

Vertex 4 gets a permanent label.

Step 2:
$P = \{1, 2, 3, 4\}$ $T = T - \{4\} = \{5, 6, 7\}$

$L(1) = 0$ $L'(5) = \text{Min} \{10, L'(4) + -\}$
$L(2) = 4$ $L'(6) = \text{Min} \{9, L'(4) + d(4, 6)\}$
$L(3) = 5$ $L'(7) = \text{Min} \{-, L'(4) + -\}$
$L(4) = 7$

Iteration 4

Step 1:
$P = \{1, 2, 3, 4\}$ $T = \{5, 6, 7\}$

$L(1) = 0$ $L'(5) = 10$
$L(2) = 4$ $L'(6) = 9$
$L(3) = 5$ $L'(7) = -$
$L(4) = 7$

Vertex 6 gets a permanent label.

Step 2:
$P = \{1, 2, 3, 4, 6\}$ $T = \{5, 6, 7\} - \{6\} = \{5, 7\}$

$$\begin{cases} L(1) = 0 \\ L(2) = 4 \\ L(3) = 5 \\ L(4) = 7 \\ L(6) = 9 \end{cases}$$

$$\begin{cases} L'(5) = \text{Min} \{10, L'(6) + d(6, 5)\} \\ L'(7) = \text{Min} \{-, L'(6) + d(6, 7)\} \end{cases}$$

Iteration 5

Step 1:
$P = \{1, 2, 3, 4, 6\}$ $T = \{5, 7\}$

$$\begin{cases} L(1) = 0 \\ L(2) = 4 \\ L(3) = 5 \\ L(4) = 7 \\ L(6) = 9 \end{cases}$$

$$\begin{cases} L'(5) = 10 \\ L'(7) = 17 \end{cases}$$

Vertex 5 gets a permanent label.

Step 2:

$P = \{1, 2, 3, 4, 6, 5\}$ $T = \{5, 7\} - \{5\}$

$\begin{cases} L(1) = 0 \\ L(2) = 4 \\ L(3) = 5 \\ L(4) = 7 \\ L(6) = 9 \\ L(5) = 10 \end{cases}$ $L'(7) = \text{Min} \{17, L'(5) + d(5, 7)\}$

Iteration 6

Step 1:

$P = \{1, 2, 3, 4, 5, 6\}$ $T = \{7\}$

$L'(7) = 16$

Vertex 7 gets a permanent label.

Step 2:

$P = \{1, 2, 3, 4, 5, 6, 7\}$ $T = $ the empty set.

$\begin{cases} L(1) = 0 \\ L(2) = 4 \\ L(3) = 5 \\ L(4) = 7 \\ L(5) = 10 \\ L(6) = 9 \\ L(7) = 16 \end{cases}$

Once we obtain the S.D. from vertex 1 to each vertex, it is very easy to determine a shortest path from 1 to each vertex. This is achieved by constructing a shortest distance tree rooted at 1 as follows.

For each vertex i (other than 1), find a vertex j such that (1) there is an arc from j to i in the network, (2) $L(j) < L(i)$, and (3) $L(j) + a(j, i) = L(i)$. Tie-breaking is arbitrary. Include arc (j, i) in the tree. In our example there are arcs from 3 and 4 to 6. We see that

$$L(3) + a(3, 6) = 5 + 4 = 9 = L(6)$$

and

$$L(4) + a(4, 6) = 7 + 5 = 12$$

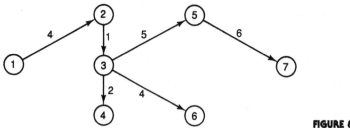

FIGURE 8.2.2

Thus arc (3, 6) is in the tree. It is easily seen that there is a tie between (3, 5) and (6, 5) to be included in the tree and we can take only one of them. In Figure 8.2.2 we have a S.D. tree rooted at 1, giving the shortest paths from 1 to all other vertices.

Note:

Dijkstra's algorithm need not solve the S.D. problem for an arbitrary weight function. Consider the network $G = (V, E)$ where $V = \{1, 2, 3\}$ and the arcs are $(1, 2)$, $(1, 3)$, and $(2, 3)$ with weights 10, 8, and -3, respectively. In iteration 2 we get $L(3) = 8$, but the S.D. from 1 to 3 is only 7.

8.3 FLOYD–WARSHALL ALGORITHM

We saw that Dijkstra's algorithm is not suitable when the weight function is arbitrary. There are several polynomial algorithms to solve the S.D. problem when the weight function is not restricted to be nonnegative. One well-known algorithm is the Floyd–Warshall algorithm, which can be used to find the S.D. and S.P. from every vertex to every other vertex for arbitrary weight functions and which locates the existence of negative cycles. If there is a negative cycle that starts at i and ends in i, it does not make sense to consider the S.D. from i to any vertex in the network in a minimization problem.

Consider a directed network with n vertices and an arbitrary weight function. Let $A = (a_{ij})$ be the $n \times n$ weight matrix and let $P = (p_{ij})$ be another $n \times n$ matrix where $p_{ij} = j$. We have n iterations during the execution of the algorithm.

Iteration j begins with two $n \times n$ matrices $A^{(j-1)}$ and $P^{(j-1)}$ (initially $A^{(0)} = A$ and $P^{(0)} = P$) and ends with $A^{(j)}$ and $P^{(j)}$. The elements in these matrices are defined as follows:

If $d_{ik}^{(j-1)} \leq d_{ij}^{(j-1)} + d_{jk}^{(j-1)}$, then the (i, k) entry in $A^{(j-1)}$ equals the (i, k) entry in $A^{(j)}$ and the (i, k) entry in $P^{(j-1)}$ equals the (i, k) entry in $P^{(j)}$. Otherwise, the (i, k) entry in $A^{(j)}$ is the sum of the (i, j) entry and the (j, k) entry in $A^{(j-1)}$ and the (i, k) entry in $P^{(j)}$ is equal to the (i, j) entry in $P^{(j-1)}$.

When the algorithm terminates we are left with two matrices, $A' = A^{(n)}$ and $P' = P^{(n)}$. It can be proved by induction that the (i, j) entry in A' is the S.D. from i to j. See Papadimitriou and Steiglitz (1982) for a proof. It can also be verified that if the (i, j) entry in P' is k, then (i, k) is the first arc in a S.P. from i to j and we use this fact to obtain a S.P. from i to j.

This procedure of updating the A-matrix (known as the **triple operation**) involves at most $(n - 1)^2$ comparisons and an equal number of additions for each iteration, whereas the procedure to update the P matrix does not involve any work. Thus the worst-case complexity is $O(n^3)$, as there are at most n iterations.

Let us illustrate this procedure in the case of the network shown in Figure 8.3.1. We compute the matrices

$$A^{(0)} = \begin{bmatrix} 0 & 4 & -3 & - \\ -3 & 0 & -7 & - \\ - & 10 & 0 & 3 \\ 5 & 6 & 6 & 0 \end{bmatrix} \qquad P^{(0)} = \begin{bmatrix} 1 & 2 & 3 & 4 \\ 1 & 2 & 3 & 4 \\ 1 & 2 & 3 & 4 \\ 1 & 2 & 3 & 4 \end{bmatrix}$$

$$A^{(1)} = \begin{bmatrix} 0 & 4 & -3 & - \\ -3 & 0 & -7 & - \\ - & 10 & 0 & 3 \\ 5 & 6 & 2 & 0 \end{bmatrix} \qquad P^{(1)} = \begin{bmatrix} 1 & 2 & 3 & 4 \\ 1 & 2 & 3 & 4 \\ 1 & 2 & 3 & 4 \\ 1 & 2 & 1 & 4 \end{bmatrix}$$

$$A^{(2)} = \begin{bmatrix} 0 & 4 & -3 & - \\ -3 & 0 & -7 & - \\ 7 & 10 & 0 & 3 \\ 3 & 6 & -1 & 0 \end{bmatrix} \qquad P^{(2)} = \begin{bmatrix} 1 & 2 & 3 & 4 \\ 1 & 2 & 3 & 4 \\ 2 & 2 & 3 & 4 \\ 2 & 2 & 2 & 4 \end{bmatrix}$$

$$A^{(3)} = \begin{bmatrix} 0 & 4 & -3 & 0 \\ -3 & 0 & -7 & -4 \\ 7 & 10 & 0 & 3 \\ 3 & 6 & -1 & 0 \end{bmatrix} \qquad P^{(3)} = \begin{bmatrix} 1 & 2 & 3 & 3 \\ 1 & 2 & 3 & 3 \\ 2 & 2 & 3 & 4 \\ 2 & 2 & 2 & 4 \end{bmatrix}$$

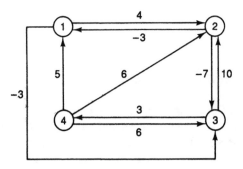

FIGURE 8.3.1

and

$$A' = A^{(4)} = \begin{bmatrix} 0 & 4 & -3 & 0 \\ -3 & 0 & -7 & -4 \\ 6 & 9 & 0 & 3 \\ 3 & 6 & -1 & 0 \end{bmatrix} \qquad P' = P^{(4)} = \begin{bmatrix} 1 & 2 & 3 & 3 \\ 1 & 2 & 3 & 3 \\ 4 & 4 & 3 & 4 \\ 2 & 2 & 2 & 4 \end{bmatrix}$$

From A' we see that the S.D. from vertex 3 to vertex 1 is the (3, 1) entry in that matrix, which is 6. From P' we see that the (3, 1) entry is 4. So (3, 4) is the first arc in the S.P. from 3 to 1. The (4, 1) entry is 2. So the next arc is (4, 2). Then we see that the (2, 1) entry is 1. Thus the last arc is (2, 1).

Next consider a digraph with four vertices for which we have

$$A = \begin{bmatrix} 0 & - & - & 1 \\ 2 & 0 & 1 & - \\ - & - & 0 & - \\ - & -4 & 3 & 0 \end{bmatrix}$$

At the end of the second iteration we have

$$A^{(2)} = \begin{bmatrix} 0 & - & - & 1 \\ 2 & 0 & 1 & 3 \\ - & - & 0 & - \\ -2 & -4 & 3 & -1 \end{bmatrix} \qquad P^{(2)} = \begin{bmatrix} 1 & 2 & 3 & 4 \\ 1 & 2 & 3 & 1 \\ 1 & 2 & 3 & 4 \\ 2 & 2 & 3 & 2 \end{bmatrix}$$

We see a new development here: *The diagonal element (4, 4) is negative* instead of being 0, indicating the existence of a negative cycle in the network. In the corresponding P matrix we see that the (4, 4) element is 2, giving the arc (4, 2); the (2, 4) element is 1, giving the arc (2, 1), and the (1, 4) element is 4, giving the arc (1, 4) creating the cycle ④- - - -②- - - - -①- - - - -④ with total weight -1.

8.4 COMPARISON OF THE TWO ALGORITHMS

To solve the all-pair (i.e., from every vertex to every other vertex) S.D. problem, it appears that on the average Dijkstra's algorithm will outperform that of Floyd–Warshall, as can be seen from the following table in Syslo et al. (1983).

Computing Times for All-Pair Shortest Path Algorithms on Complete Networks

Number of vertices	Run time (sec)	
	Dijkstra	Floyd–Warshall
40	0.527	0.646
60	1.767	2.156
80	4.208	5.078
100	8.052	9.862

8.5 NOTES AND REFERENCES

No other problem in network optimization has received as much attention as the shortest distance problem. For an excellent review, see Dreyfus (1969). Some excellent general references are the relevant chapters in the books by Lawler (1976), Minieka (1978), and Papadimitriou and Steiglitz (1982). The paper by Dijkstra (1959) is one of the earliest papers on this topic. See Nemhauser (1972) for an extension of Dijkstra's algorithm for networks with arbitrary weights. The Floyd–Warshall algorithm was published as an ALGOL algorithm by Floyd (1964) based on the work of Warshall (1962). This algorithm is by far one of the most efficient known algorithms for solving the all-pair shortest distance problem. For another efficient algorithm, see Tabourier (1973).

8.6 EXERCISES

8.1. The distance matrix of a digraph is as follows:

$$
\begin{bmatrix}
0 & - & 4 & 10 & 3 & - & - \\
- & 0 & -1 & -1 & 2 & 11 & 0 \\
- & 9 & 0 & 8 & 3 & 2 & 1 \\
- & 4 & 0 & 0 & 8 & 6 & 3 \\
- & 0 & 1 & 2 & 0 & 3 & -1 \\
- & -1 & -1 & 3 & 2 & 0 & 0 \\
- & 4 & 3 & - & - & 2 & 0
\end{bmatrix}
$$

Find $A^{(7)}$ and $P^{(7)}$ using the Floyd–Warshall algorithm.

8.2. Find the S.D. and a S.P. from vertex 4 to vertex 7 in Problem 8.1.

8.3. Construct a directed tree rooted at vertex 1 giving the S.D. from 1 to the other vertices in problem 8.1.

8.4. Replace the number -1 that appears in the fourth column of the matrix in Problem 8.1 by -3. You detect a negative cycle now. What is this negative cycle?

8.5. Find a S.P. from 4 to 2 that does not pass through 5, 6, or 7 in problem 8.1.

8.6. Replace -1 by 1 in the matrix of Problem 8.1 and find a tree rooted at vertex 1 giving the S.D. from vertex 1 to all vertices using Dijkstra's algorithm.

8.7. At a small but growing airport the local airline company is purchasing a new tractor–trailer train to bring luggage to and from the airplanes. A new mechanized luggage system will be installed in three years and the tractor will not be needed after that. However, it may be more economical to replace the tractor after one or two years because, due to heavy use, running time and maintenance cost will increase rapidly with age. The following array gives the total net discounted cost associated with purchasing a tractor (purchase price minus trade-in allowance plus running and maintenance costs) at the end of year i and trading it in at the end of year j (assuming that year 0 is now).

$$
\begin{array}{c}
\quad\quad\quad\ j \\
\quad\quad 1 \quad\ 2 \quad\ 3 \\
\begin{array}{cc}
 & 0 \\
i & 1 \\
 & 2
\end{array}
\left[
\begin{array}{ccc}
12 & 27 & 47 \\
   & 15 & 32 \\
   &   & 18
\end{array}
\right]
\end{array}
$$

The problem is to determine at what times (if any) the tractor should be replaced to minimize the total cost. Formulate this as a shortest distance problem and solve it. (This problem is from Hillier and Lieberman, 1986.)

8.8. A garage sells used motorcycles for $500.00 each. The sale is only at the beginning of the academic year and the purchase price is the same every year. A student can buy a vehicle and use it for four years or replace it after using it for one year or two years or three years. The trade-in value of a vehicle is $100.00 after one year, $50.00 after two years, $30.00 after three years, and $0.00 after four years. The maintenance costs for a used vehicle are $200.00, $400.00, $600.00, and $800.00, respectively, for the four years. The problem for a student interested in buying a motorcycle from the dealer for use for four years in college is to make a decision as to whether the cycle bought as a freshman should be kept for all four years or it should be replaced so that the total cost is a minimum. If replacement is necessary, at what intervals? Formulate this problem as a shortest distance problem and solve it.

8.9. If the weights of the edges of a connected graph are all distinct positive numbers, is it true that there is a unique minimal spanning tree in the graph? Is it true that the shortest path between any two vertices is unique?

8.10. A network is said to satisfy the triangle inequality if for every three distinct arcs (i, j), (j, k), and (i, k) the weight of the arc (i, k) does not exceed the sum of the weights of the other two arcs. A directed path P from vertex x to vertex y is a minimum length path if the number of arcs in P does not exceed the number of arcs in any other path from x to y. Is it necessary that the shortest path from x to y is a minimum length path?

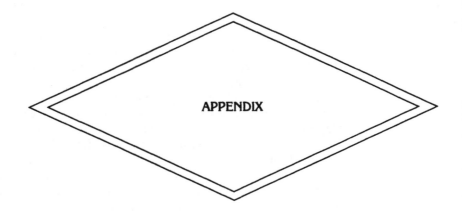

APPENDIX

What Is
NP-Completeness?

A.1 PROBLEMS AND THEIR INSTANCES

Informally, we always distinguish between a *problem* and an *instance* of the problem. For example, "solve a linear system of equations" is a problem an instance of which will be "given a $m \times n$ matrix A and a $m \times 1$ matrix B test whether there exists a $n \times 1$ matrix x such that $Ax = B$ and if the answer is yes obtain the vector x." Here A and B are the "inputs" and x is the "solution" or "output." On the other hand, an instance of the *decision problem* "does a linear system have a solution?" will be "is there a x such that $Ax = B$?" The output now is either "yes" or "no." Problems and decision problems thus viewed have an infinite number of instances.

A more formal approach to the concept of " problem" and "an instance of the problem" as defined in Schrijver (1986) is along the following lines. An **alphabet** is a finite set L the elements of which are **symbols** or **letters.** An ordered finite sequence of symbols from L is called a **string** or a **word.** The set of all strings or words from L is denoted by L^*. The size or **length** of a string is the number of components in it. If $L = \{a, b, c\}$, the length of $x = abbaa$ is 5 even though x consists of only two symbols. The string of size 0 is called the **empty string,** denoted by ϕ.

There are several ways of encoding rational numbers, vectors, matrices, systems of inequalities, linear systems, graphs (matrix representations), and so on, as strings of symbols from a fixed alphabet such as $L = \{0, 1\}$. In our present discussion we shall skip the details of such encodings. For more details, see Garey and Johnson (1979).

A **problem** p is a subset of $L^* \times L^*$. For any problem p we have the corresponding metamathematical problem: Given a string z in L^*, find a string y in L^* such that (z, y) is in p or report that no such string y exists. Here the string z is called an **instance** or **input** of the problem and y is called the **solution** or **output.** A problem p is called a **decision problem** if whenever (z, y) is in p, then y is the empty string. If $L^*(p) = \{z \in L^* : (z, \phi) \in p\}$, the corresponding metamathematical problem is: Given a string z in L^*, does it belong to $L^*(p)$?

Thus the set $\{((A, B), x) : A$ is a matrix, B and x are column vectors such that $Ax = B\}$ is a subset of $L^* \times L^*$ (where $L = \{0, 1\}$), defining the problem p for which (A, B) is an instance and x is a solution. This problem can be couched in metamathematical language as follows: Given the string (A, B), find a string x (if it exists) such that $Ax = B$. Similarly, the set $\{((A, B), \phi) : A$ is a matrix, B is a column vector such that $Ax = B$ for at least one column vector $x\}$ is a decision problem the metamathematical version of which is as follows: Given a matrix A and a column vector b, is there a column vector x such that $Ax = B$?

A.2 THE SIZE OF AN INSTANCE

If the rational number q is of the form m/n (where m and n are relatively prime, m is an integer, and n is a positive integer), the size of q is defined to be size $(q) = 1 + $ ceiling of $[\log(m + 1)] + $ ceiling of $[\log(n + 1)]$. There are other ways of defining the size of a rational number. But it can be shown that most of these definitions are "linearly equivalent" (see Section A.5 for a definition of linear equivalence). Thus the size of a positive integer is proportional to its logarithm and not to its value. See Garey and Johnson (1979) again for sizes of instances and encodings. If A is a vector with n rational components, we define size $(A) = n + $ the sum of the sizes of all the components of A. Similarly, if M is a $m \times n$ rational matrix, define size $(M) = mn + $ the sum of the sizes of all the elements of the matrix. The size of the linear equation $ax = b$ or the linear inequality $ax \le b$ is $1 + $ size$(a) + $ size(b). The size of the linear system $Ax = B$ is $1 + $ size$(A) + $ size(B). The size of a graph is the size of its incidence matrix.

A.3 ALGORITHM TO SOLVE A PROBLEM

An **algorithm** to solve a problem, in an informal sense, is a finite sequence of instructions to obtain an output for a given input of the problem. It is a step-by-step procedure for solving the problem. Thus given the instance z in L^*, an algorithm for the problem p determines an output y in L^* such that (z, y) is in p or terminates without delivering an output if there is no such string y. It is possible to define an algorithm in a more formal sense in terms of Turing machines or computer programs in some programming language. For our purpose this informal concept of an algorithm will be sufficient. We mention in passing that there are well-defined problems in mathematics for which no algorithms exist. A problem is **undecidable** if there is no algorithm that will solve every instance of the problem. It was proved in 1970 by the then 22-year-old Russian mathematician Yuri Matiyasevich that the decision problem known as **Hilbert's tenth problem,** which asks whether a polynomial equation in more than one variable with integer coefficients has any integer solutions, is an undecidable problem. The most famous undecidable problem in computer science is the **halting problem:** Given a computer program with its input, will it ever halt? When we say that the halting problem is undecidable, we mean that there is no algorithm which will decide whether an arbitrary computer program will get into an infinite loop while working on a given input. An excellent reference for the topic of undecidability and related items of interest is the book by Lewis and Papadimitriou (1981).

A.4 COMPLEXITY OF AN ALGORITHM

If we have two algorithms at our disposal to solve every instance of a problem, it is natural to compare them to find out whether one is better or more efficient than the other. For this purpose we have to measure the amount of work done by the algorithm, which is the number of "basic operations" needed to solve the problem using the algorithm. Here are some examples of basic operations. The basic operation in a sorting problem is the comparison of two numbers in a given list of entries and thus the work done in a sorting problem is the number of comparisons. In a problem involving both multiplication and addition, we may take multiplication as a basic operation since multiplication is more difficult than addition. Thus the work done in multiplying two $n \times n$ matrices is at most n^3. (See Section 3.5.)

If A is an algorithm to solve (every instance of) a problem and if x is an instance of the problem, the number of basic operations needed to solve x using A is denoted by $w_A(x)$. Since we are interested in the efficiency of the actual working of the algorithm, it is important that this measure we choose to define the work done is independent of the computer used, the particular computer program, programming language, and other implementation details. Usually, the work done is taken as a function of the *size* of an instance of the problem. Now if two instances have the same size, it does not imply that the work done is the same for both. Thus we have to aggregate in one way or other the work done *for* all instances of the same length.

One way of doing this is by taking a worst-case approach. Thus the **worst-case complexity** of the algorithm A to solve the problem p is defined to be $f_A(n)$, where $f_A(n) = \text{Max } \{w_A(x) : x$ is an instance of p and size of x is $n\}$ We may take a different approach as follows. Let $h(x)$ be the probability that an instance x of the problem is taken as a candidate for input. Then the **average-case complexity** is the sum of all terms $h(x) \cdot w_A(x)$ where x is an instance of size n. In what follows complexity means complexity in the worst case.

A.5 THE "BIG OH" OR THE O(·) NOTATION

Let f and g be two functions from the set of natural numbers to the set of nonnegative real numbers. If there is a positive constant c and a natural number n_0 such that $f(n) \le c \cdot g(n)$ for all $n \ge n_0$, we write "f is $O(g)$" or "$f = O(g)$" or "$f(n)$ is $O(g(n))$" or "$f(n) = O(g(n))$" and say (as in Wilf, 1986) that "$f(n)$ **is big oh of** $g(n)$." Two functions f and g are **linearly equivalent** if $f = O(g)$ and $g = O(f)$. We write $f >< g$ if f and g are linearly equivalent. The relation $><$ thus defined is an equivalence relation and the **rate of growth** of f is its equivalence class, which may be represented by a canonical member from that class. For example, let $f(n) = 5n^2 + 9n + 7$ and $g(n) = 8n^2 + 23$. Then it is easy to show that $f >< g$ and a typical representative from the equivalence class to which f and g belong is the function $p(n) = n^2$. Thus we write $f(n) = O(n^2)$ and $g(n) = O(n^2)$. Now consider the function $h(n) = 4n^3 + 9n$. Then $f(n)$ is $O(h(n))$ but $h(n)$ is definitely not of $O(f(n))$. Notice that the big oh notation gives only an upper limit. If $f(n)$ is $O(n^k)$, it is quite possible that $f(n)$ is $O(n^r)$ for some r less than k. At the same time $f(n)$ is $O(n^r)$ for all $r \ge k$.

Let $c = \lim [f(n)/g(n)]$ as n goes to plus infinity. Then it can easily be verified that:

1. If c is finite and nonzero, f and g are linearly equivalent.
2. If c is zero, f is $O(g)$ but g is not $O(f)$.
3. If c is infinite, g is $O(f)$ but f is not $O(g)$.

As a consequence, we say that f is of **lower order** than g (or equivalently, g is of **higher order** than f) if $c = 0$. Using this ratio test one can establish that if k is a positive integer, n^k is of higher order than $\log n$, 2^n is of higher order than n^k, and $n!$ is of higher order than 2^n.

A.6 EASY PROBLEMS AND DIFFICULT PROBLEMS

An algorithm A to solve a problem p is called a **polynomial algorithm** or **polynomial-time algorithm** if its worst-case complexity $f_A(n)$ is $O(n^k)$ for some fixed positive integer k. Thus an algorithm with complexity $n \log n$ is a polynomial algorithm because $n \log n$ is $O(n^2)$. An algorithm whose complexity violates all polynomial bounds is referred to as an **exponential algorithm.** An algorithm with complexity $f(n)$ is exponential if and only if there are positive numbers a and b, numbers p and q greater than 1, and a positive integer n_0 such that $a \cdot p^n \le f(n) \le b \cdot q^n$ for all $n \ge n_0$. Some examples of rates of growth of exponential algorithms are k^n $(k > 1)$, $n!$, n^n, and $n^{\log n}$. To discuss whether a problem is easy or not, it should first be decided where to draw a line between easy and difficult problems. The distinction between exponential functions and polynomials becomes clear if we take an asymptotic point of view: Polynomials grow more slowly than exponential functions. So polynomial algorithms with growth rate n^k (even when k is large) are efficient in comparison with exponential functions. In other words, for sufficiently large problems, a polynomial algorithm executed on the slowest computer will solve a problem faster than an exponential algorithm on the fastest computer. Furthermore, in some cases an algorithm to solve a problem may be obtained by combining several algorithms for simpler subproblems. If each of these algorithms of subproblems is polynomial, then the algorithm of the main problem is also polynomial because the class of all polynomials is closed under addition, multiplication, and composition of functions. Thus the consensus among computer scientists is to say that a problem is **easy** if there is a polynomial algorithm that will solve every instance of the problem. This idea is originally due to Edmonds (1965b). Observe that it does not make sense to say that an algorithm is good if its complexity is $O(n^k)$ when k is large. In this connection the following comments from Papadimitriou and Steiglitz (1982) are worth reproducing: "The thesis that polynomial-time algorithms are 'good' seems to weaken when pushed to extremes. Experience, however, comes to its support. For most problems, once any polynomial-time algorithm is discovered, the degree of the polynomial quickly undergoes a series of decrements as various researchers improve on the idea. Usually the final rate of growth is $O(n^3)$ or better."

To appreciate the difference in the rate of growth of a polynomial algorithm and the tyrannical rate of growth of an exponential algorithm, consider the following scenario: Suppose that a basic operation (each step) in a computer requires one millionth of a second of computer time. If $n = 50$, the computation times for algorithms with complexities n^2, n^3, 2^n, and 3^n will be 0.0025 second, 0.125 second, 35.7 years, and $(2)(10^9)$ centuries, respectively. If $n = 100$, these times will be 0.01 second, 1 second, 2^{48} centuries, and 3^{70} centuries, respectively.

It was mentioned earlier that a problem is (provably) undecidable if it can be proved that there exists no algorithm that will solve every instance of the problem. Instead of asking whether a problem is provably undecidable or not, the basic question in computational complexity theory asks how difficult or hard it is to solve a problem. A problem is **provably difficult** if it can be proved that any algorithm which will solve (every instance of) the problem is an exponential algorithm. Such problems do exist, but they are rather obscure. See Lewis and Papadimitriou (1981) for more details. A problem for which no polynomial algorithm is known and for which it is conjectured that no such algorithm exists is called an **intractable** problem.

The problem of finding a shortest path between a vertex and another vertex in a connected graph is an easy problem (the algorithm of Dijkstra is polynomial), but the problem of finding a longest simple path between two vertices is intractable since no one knows an algorithm to solve this which is substantially faster than enumerating all possible paths between the two and choosing the optimal one.

Thus if there is a polynomial algorithm to solve a problem p, we can say that "it is easy to solve p." But how can one show that "it is hard (not easy) to solve a problem"? One may prove that problem p is as hard as problem q. So whenever there is no known polynomial algorithm to solve a problem, we have to examine that problem in this spirit. In other words, to prove that a given problem is difficult, it is not enough to assert that no polynomial algorithm has been discovered so far to solve it.

It requires some sophisticated mathematical techniques to show that the complexity of any algorithm conceivable for the problem cannot be bounded above by a polynomial. Such techniques are now being discovered by computer scientists with the advent of latest developments in computational complexity theory. The introduction of the concept of NP-completeness is an important milestone in this field.

A.7 THE CLASS P AND THE CLASS NP

Hereafter we shall assume that the problems we consider are all decision problems. These are problems whose output is either "yes" or "no." Examples of decision problems: (1) Is there a Hamiltonian cycle in a given connected graph? and (2) Is there a simple path from a vertex v to another vertex w in a connected network such that the total length of this path does not exceed a certain amount?

We are interested in classifying decision problems according to their complexity. A decision problem belongs to **class P** if there is a polynomial algorithm to solve every instance of the problem. If a problem has a polynomial algorithm, then obviously the corresponding decision problem also has a polynomial algorithm.

Notice that we can assert that an arbitrary problem is in the class P only after we have a proof that there is a polynomial algorithm to solve it. The fact that the linear programming problem is a member of the class P was established only a decade ago, when Khachiyan (1979) came out with his ellipsoid algorithm. Subsequently, a more efficient algorithm was obtained by Karmarkar (1984) using interior point methods. For a lucid description of these developments, see Schrijver (1986).

Now consider a decision problem the status of which is as follows: (1) there is at least one exponential algorithm to solve it, and (2) no one has proved so far that every algorithm to solve is necessarily exponential. In other words, it is a decidable decision problem and we do not know whether or not it is provably difficult. What do we do with problems in this category?

It is at this stage that we introduce a class of decision problems containing the class P. A decision problem belongs to **class NP** if there is a polynomial algorithm to verify the "yes" output of that problem. The acronym NP is for "nondeterministic polynomial." For details regarding nondeterministic algorithms, see Garey and Johnson (1979). Nondeterministic algorithms are not in any sense probabilistic or random algorithms.

It is obvious that any decision problem with a polynomial algorithm is in the class NP. Regarding a typical problem p in NP for which so far no one has obtained a polynomial algorithm, there are three mutually exclusive alternatives: (1) a polynomial algorithm for p will be discovered, (2) it will be proved that p will not have a polynomial algorithm, and (3) the status of p will never be settled.

So far no one has proved that there exists a problem in NP that is not in P. it is not known whether P = NP or whether P is properly contained in NP. This is a frustrating situation because many practical problems belong to the class NP. If P = NP, it makes sense to try to obtain a polynomial algorithm for a problem in NP for which no efficient algorithm has been discovered so far. On the other hand, if we could establish that a particular problem in NP is not in P, we need not bother to seek an efficient algorithm to solve it. But in the absence of a proof, we cannot abandon our efforts to obtain a polynomial algorithm to solve this problem because there is always a remote chance that somewhere out there a polynomial algorithm to solve the problem exists waiting to be discovered!

Here is an example of a problem in NP. Consider the decision problem an instance of which is as follows: "Is the positive integer n a composite number?" When n is large, we cannot easily answer this question. However, if we are able to exhibit two positive integers p and q (both greater than 1) such that $n = pq$, then anyone can easily say the answer is "yes" since multiplication of two numbers can be performed in polynomial time. On the other hand, it is not at all obvious whether the decision problem "is the positive integer n a prime number?" belongs to the class NP. It was proved by Pratt (1975) that this is indeed the case.

Corresponding to each decision problem p, there is always a complementary decision problem p'. Each is complementary to the other. The problems "is it true that n is prime?" and "is it true that n is not prime?" are complementary. In a connected graph G the problems "is it true that there is a Hamiltonian cycle in G?" and "is it true that there is no Hamiltonian cycle in G?" are complementary. Notice that the former problem in graph theory is in NP because it is easy to verify the "yes" answer. But the latter problem is not in NP because to verify the "yes" answer in this case, we have to enumerate all possible cycles. Incidentally, we have here a decision problem that is not in NP. Thus if a problem is in NP it is not necessary that its complementary problem is in NP. A decision problem belongs to **class Co-NP** if its complementary problem is in class NP. A decision problem is said to be **well-characterized** if it is in both NP and Co-NP. Obviously, any problem in P is well-characterized. It is not known whether every well-characterized problem is in P. The problems "is n prime?" and "is n composite?" are both well-characterized because of Pratt's theorem.

It is also not known whether NP = Co-NP. If NP is contained in Co-NP (or if Co-NP is contained in NP), then NP = Co-NP = NP $\cap$ Co-NP. Also, if P = NP, all the sets coincide.

A.8 POLYNOMIAL TRANSFORMATIONS AND NP-COMPLETENESS

The easy problems in NP are in P. They are on one side of the spectrum. Since it is not known whether P = NP or not, it is natural to ask whether we can collect all the "hard" problems of NP in one class and put this class on the other side of the spectrum. In the first place, what should be the name of this distinguished class? It seems that Donald Knuth of Stanford University polled his colleagues in 1974 to find an appropriate name. There were many suggestions: formidable, Herculean, arduous, prodigious, obstinate, and so on. Finally, a consensus: Call it the class of NP-complete problems—and this name stuck among computer scientists, logicians, and mathematicians.

A basic idea in the theory of NP-completeness is that of polynomial transformation. A decision problem p is **polynomially transformable** to a decision problem q if the following two conditions hold: (1) there exists a function $f(x)$ that will transform every instance x of p to an instance $f(x)$ of q such that the answer to x is "yes" if and only if the answer to $f(x)$ is "yes" and (2) there is an efficient algorithm to compute $f(x)$ for every x.

Here is an example of a problem that can be polynomially transformed into another. Recall that a clique in a graph G is a complete subgraph of G. The number of vertices in a clique is its size. The **clique problem** p is stated as follows: Is there a clique of a prescribed size in a graph? For a given finite set, a collection of subsets is said to cover the set if every element in the set belongs to at least one set in the collection. The **set covering problem** q is stated as follows: Given a finite set X, a collection C of subsets of X, and a positive integer m, is there a subcollection C' consisting of m of these sets such that C' covers X?

The problem p can be transformed into q as follows. Let $G = (V, E)$ be a connected graph with m edges where $V = \{1, 2, 3, \ldots, n\}$. Then its complement $G' = (V, E')$ has r edges, where $r = [n(n - 1)/2] - m$. Suppose that $E' = \{e_1, e_2, \ldots, e_r\}$ is the set to be covered. Let S_i be the set of edges in G' that are incident at vertex i. We take $C = \{S_i : i = 1, 2, \ldots, n\}$ as a collection of available subsets of E'. It is easy to see that if G has a clique of size k, then a subcollection C' of $(n - k)$ subsets can be chosen from C that will cover the set E'. In particular, if $W = \{1, 2, \ldots, k\}$ is a set of vertices that forms a clique in G, the collection $C' = \{S_{k+1}, S_{k+2}, \ldots, S_n\}$ will be a cover for E'.

The definition of polynomial transformability suggests the following inequality: If p is polynomially transformable to q, (complexity of p) $\leq$ (complexity of f) + (complexity of q). Furthermore, if complexity of f is insignificant compared to the complexities of p and q, we can write, in an asymptotic sense, (complexity of p) $\leq$ (complexity of q).

A decision problem p is **NP-hard** if every problem in NP can be transformed into it polynomially. In other words, an NP-hard problem cannot be easier than any problem in NP. A problem in NP that is NP-hard is said to be **NP-complete.** The class of NP-complete problems is thus the intersection of the class NP and and class NP-hard. Thus if there is an efficient algorithm to solve every instance of a particular NP-complete problem, every problem in NP has a polynomial algorithm. The class of NP- complete problems is denoted by NPC. To show that a problem p is in NPC it has to be proved that (1) p is in NP and (2) every problem in NP can be polynomially transformed into p. Observe that the complexity of a problem in NPC is inextricably related to the conjecture that P is a proper subset of NP.

The fact that the class NPC is nonempty was established by Cook (1971) in his seminal paper by exhibiting a problem in NP such that every problem in NP can be transformed into it polynomially. This problem is known as the **satisfiability problem.** This problem comes from mathematical logic and applications in switching theory. However, it can be stated as a simple combinatorial puzzle as in Karp (1986):

> Given several sequences of upper- and lowercase letters, is it possible to select a letter from each sequence without selecting both the upper- and lowercase versions of the same sequence? For example, if the sequences are Abc, BC, aB, and ac, it is possible to choose A from the first sequence, B from the second and third, and c from the fourth; note that the same letter can be chosen more than once, provided we do not choose both its uppercase and lowercase versions. An example where there is no way to make the required selections is given by the four sequences AB, Ab, aB and ab.

The satisfiability problem is clearly in NP, since it is easy to check that whether a proposed selection of letters satisfies the conditions of the problem. Cook proved that, if the satisfiability problem is solvable in polynomial time, then every problem in NP is solvable in polynomial time, so that $P = NP$. Thus we see that this seemingly bizarre and inconsequential problem is an archetypal combinatorial problem; it holds the key to the efficient solutions of all problems in NP.

The "floodgate" was open once it was proved that the class NPC is nonempty. By constructing a series of polynomial transformations Karp (1972a) produced a list of 20 or so problems in NPC. It was shown in this paper that most of the classical combinatorial problems such as packing, covering, matching, partitioning, routing, and so on, are in NPC. His list includes the following: (1) Is a given graph Hamiltonian?; (2) Is it possible to color the vertices of a graph with k colors so that no two adjacent vertices have the same color?; (3) Given a set of numbers $\{n_i : i = 1, 2, \ldots, k\}$ and a number s, does some subset of the numbers add up to exactly s?; and (4) Is there a clique of a given size in a graph?

If p is a problem in NP, to show that p is in NPC it is enough if we prove that some known problem in NPC is polynomially transformable to it. There are thousands of problems now known to be NP-complete and their "tribe" is steadily increasing, as can be seen from the publications of new results in this field in recent years. For a fascinating account of the world of NP-completeness, one shoud refer to the book by Garey and Johnson (1979) and Johnson's column on this topic entitled "NP-Completeness: An Ongoing Guide," which appears regularly in the *Journal of Algorithms*. It is now routine to investigate whether an apparently difficult problem is NP-complete.

Recall that a problem in NP is well-characterized if its complement also is in NP. If it can be proved that there exists a problem in NPC that is well-characterized, it can be shown that NP = Co-NP. Attempts to show that complements of some standard NP-complete problems are in NP have been fruitless. Also, there is no evidence to believe that these two classes NP and Co-NP coincide. Hence the conjecture: The class NPC and the class W of well-characterized problems are disjoint. (It was known for some time that the linear programming problem LP is a well-characterized problem. So it was conjectured that LP is not in NPC even before the discovery of the ellipsoid algorithm, which proved that not only that LP is not in NPC but that it is in P.) Similarly, if Co-NPC is the class of complements of all NP-complete problems, the class Co-NPC and the class W are also disjoint. Thus the conjectured topography of the class of decision problems is as portrayed in Figure A.8.1.

Finally, even if we *assume* that every NP-complete problem is provably difficult, there remains a class of problems of unsettled status. For example, consider a decision problem p in NP such that (1) no polynomial algorithm has

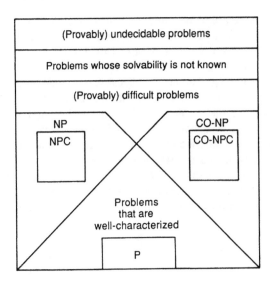

FIGURE A.8.1

been obtained so far to solve every instance of p, and (2) so far there is no proof that p is in NPC. In particular, the status of a well-characterized problem (which is not likely to be in NPC) for which no polynomial algorithm has been obtained so far remains unsettled. A typical member of this class: "Is the given positive integer a prime number?"

A.9 COPING WITH HARD PROBLEMS

The problems considered thus far are decision problems. In a *combinatorial optimization problem* there may be many solutions (feasible solutions) and each solution will have a real number associated with it called the *value* of the solution. The aim of the problem is to obtain a solution (optimal feasible solution) whose value is optimum. Corresponding to any such optimization problem, there is a decision problem of determining whether the optimization problem has a solution with a value better than a given real number. Obviously, the decision problem associated with a combinatorial optimization problem cannot be harder than the optimization problem itself. Thus if the problem "Is a graph Hamiltonian?" is hard, then the problem "find an optimal Hamiltonian cycle in a graph" is also hard. Stated more precisely, this means that if the decision problem associated with an optimization problem is NP-complete, the optimization problem is NP-hard.

Now a proposition which proves that the decision problem (which corresponds to an optimization problem) is in NPC eliminates for all practical purposes the possibility of obtaining an efficient algorithm to solve every instance of the optimization problem. But the fact remains that many combinatorial optimization problems which arise in several areas in science, engineering, and operations research are NP-hard. So it is natural to ask: How do we cope with such hard problems? Broadly speaking, there are two approaches. One is a *heuristic* approach: It is likely that the problem is hard because of a small proportion of hard instances. Is it possible to obtain an efficient algorithm that will solve a large number of instances of the problem? A heuristic algorithm always gives an optimal solution, but it need not be efficient in every instance. The simplex method to solve the linear programming problem is an algorithm of this category. The other approach is to find out whether there is an efficient *approximation algorithm* to obtain a feasible solution with value that is very close to the optimal value. There are efficient approximation algorithms for some well-known NP-hard problems in combinatorial optimization. Once again, see Garey and Johnson (1979) for more details.

Bibliography

AHO, A. V., HOPCROFT, J. E., and ULLMAN, J. D. *Data Structures and Algorithms*, Addison-Wesley, Reading, Mass., 1983.

AIGNER, M. *Combinatorial Theory*, Springer-Verlag, New York, 1979.

ANDERSON, I. *A First Course in Combinatorial Mathematics*, Oxford University Press, Oxford, 1979.

APPEL, K., and HAKEN, W. "Every Planar Map Is Four Colorable," *Bull. Amer. Math. Soc.* 82 (1976), 711–712.

BAASE, S. *Computer Algorithms: Introduction to Design and Analysis*, Addison-Wesley, Reading, Mass., 1978.

BEHZAD, M., CHATRAND, G., and LESINAK-FOSTER, L. *Graphs and Digraphs*, Wadsworth, Belmont, Calif., 1979.

BELLMORE, M., and NEMHAUSER, G. L. "The Traveling Salesman Problem: A Survey," *Oper. Res.* 16 (1968), 538–558.

BERGE, C. *The Theory of Graphs and Its Applications*, Wiley, New York, 1962.

BIRKHOFF, G. D., and LEWIS, D. C. "Chromatic Polynomials," *Trans. Amer. Math. Soc.* 60 (1960), 355–451.

BONDY, J. A., and MURTY, U. S. R. *Graph Theory with Applications*, Elsevier, New York, 1976.

BOYER, C. B. *History of Mathematics*, Wiley, New York, 1968.

BUSSEY, W. H. "Origins of Mathematical Induction," *American Math. Monthly* 24(1917), 199–207.

CARRE, B. *Graphs and Networks*, Clarendon Press, Oxford, 1979.

CHANG, S. K. "The Generation of Minimal Trees in a Steiner Topology," *J. Assoc. Comput. Mach.* 19 (1972), 699–711.

CHARTRAND, G. *Graphs as Mathematical Models*, Wadsworth, Belmont, Calif., 1977.

CHARTRAND, G., KAPOOR, S. F., and KRONK, H. V. "A Generalization of Hamiltonian-Connected Graphs," *J. Math. Pures Appl.* (9) 48 (1969), 109–116.

CHERITON, D., and TARJAN, R. E. "Finding Minimum Spanning Trees," *SIAM. J. Comput.* 5 (1976), 724–742.

COHEN, D. I. *Basic Techniques of Combinatorial Theory*, Wiley, New York, 1978.

COOK, S. A. "The Complexity of Theorem-Proving Procedures," in *Proceedings Third ACM Symposium on the Theory of Computing*, Assoc. for Computing Machinery, New York, 1971, pp. 151–158.

DE BRUIJN, N. G. "A Combinatorial Problem," *Nederl. Akad. Wetensch. Proc.* 49 (1946), 758–764.

DEO, N. *Graph Theory with Applications to Engineering and Computer Science*, Prentice-Hall, Englewood Cliffs, N.J., 1974.

DIJKSTRA, E. W. "A Note on Two Problems in Connection with Graphs," *Numer. Math.* 1 (1959), 269–271.

DIRAC, G. A. "Some Theorems on Abstract Graphs," *Proc. London Math. Soc.* 2 (1952), 69–81.

DREYFUS, S. E. "An Appraisal of Some Shortest-Path Algorithms," *Oper. Res.* 17 (1969), 395–412.

EDMONDS, J. "Paths, Trees and Flowers," *Canad. J. Math.* 17 (1965b), 449–467.

EVEN, S. *Graph Algorithms*, Computer Science Press, Potomac, Md., 1979.

FLOYD, R. W. "Algorithm 97: Shortest Path," *Comm. ACM* 7 (1964), 345.

GABOW, H. P., GALIL, Z., SPENCER, T., and TARJAN, R. E. "Efficient Algorithms for Finding Minimum Spanning Trees in Undirected and Directed Graphs," *Combinatorica* 6 (1986), 109–112.

GAREY, M. R., and JOHNSON, D. S. *Computers and Intractability: A Guide to the Theory of NP-Completeness*, W. H. Freeman, San Francisco, 1979.

GIBBONS, A. *Algorithmic Graph Theory*, Cambridge University Press, Cambridge, 1985.

GOLDBERG, S. *Introduction to Difference Equations*, Wiley, New York, 1958.

GOLDMAN, A. J. "Discrete Mathematics in Government," Lecture on the Applications of Discrete Mathematics, SIAM, Troy, N.Y. (1982).

GOLOMB, S. W. *Shift Register Sequences*, Holden-Day, San Francisco, 1967.

GOLOVINA, L. I., and YAGLOM, I. M. *Induction in Geometry*, D. C. Heath, Boston, 1963.

GONDRAN, M., and MINOUX, M. *Graphs and Algorithms*, Wiley, New York, 1984.

GOULD, R. *Graph Theory*, Benjamin-Cummings, Menlo Park, Calif., 1988.

GRAHAM R. L., and HELL, P. "On the History of the Minimum Spanning Tree Problem," *Bell Lab. Rep.* (1982).

GRAHAM, R. L., ROTHSCHILD, B. L., and SPENCER, J. H. *Ramsey Theory*, Wiley, New York, 1980.

GRIMALDI, R. P. *Discrete and Combinatorial Mathematics*, Addison-Wesley, Reading, Mass., 1985.

HAKEN, W. "An Attempt to Understand the Four Color Problem," *J. Graph Theory* 1 (1977), 193–206.

HALMOS, P. *Naive Set Theory*, Van Nostrand, Princeton, N.J., 1960.

HARARY, F. *Graph Theory*, Addison-Wesley, Reading, Mass., 1969a.

HARARY, F. "The Four Color Conjecture and Other Graphical Diseases," in *Proof Techniques in Graph Theory*, Academic Press, New York, 1969b.

HENKIN, L. "On Mathematical Induction," *Amer. Math. Monthly* 67 (1960), 323–337.

HILLIER, F. S., and LIEBERMAN, G. J. *Introduction to Operations Research*, 4th ed., Holden-Day, Oakland, Calif., 1986.

HU, T. C. *Combinatorial Algorithms*, Addison-Wesley, Reading, Mass., 1982.

HUFFMAN, D. A. "A Method for the Construction of Minimum Redundancy Codes," *Proc. IRE* 40 (1952), 1098–1101.

HUTCHINSON, J. P., and WILF, H. S. "On Eulerian Circuits and Words with Prescribed Adjacency Patterns," *J. Combin. Theory* A18 (1975), 80–87.

KARMARKAR, N. "A New Polynomial Algorithm for Linear Programming," *Combinatorica* 4 (1984), 373–395.

KARP, R. M. "Reducibility among Combinatorial Problems," in *Complexity of Computer Computations*, Plenum Press, New York, 1972a.

KARP, R. M. "A Simple Derivation of Edmonds' Algorithm for Optimum Branchings," *Networks* 1 (1972b), 265–272.

KARP, R. M. "Combinatorics, Complexity and Randomness," *Comm. ACM* 29 (1986), 98–111.

KHACHIYAN, L. G. "A Polynomial Algorithm in Linear Programming" (in Russian); English translation in *Soviet Math. Dokl.* 20 (1979), 191–194.

KNUTH, D. E. *The Art of Computer Programming*, Vol. 1, Addison-Wesley, Reading, Mass., 1973a.

KNUTH, D. E. *The Art of Computer Programming*, Vol. 3, Addison-Wesley, Reading, Mass., 1973b.

KRISHNAMURTHY, V. *Combinatorics: Theory and Applications*, Ellis Horwood, Chichester, West Sussex, England, 1986.

KRUSKAL, J. B. "On the Shortest Spanning Subtree of a Graph and the Traveling Salesman Problem," *Proc. Amer. Math. Soc.* 7 (1956), 48–50.

KWAN, M. K. "Graphic Programming Using Odd or Even Points," Chinese J. Math. 1 (1962), 273–277.

LAWLER, E. L. *Combinatorial Optimization: Networks and Matroids*, Holt, Rinehart and Winston, New York, 1976.

LAWLER, E. L., LENSTRA, J. K., RINNOOY KAN, A. H. G., and SHMOYS, D. B. *The Traveling Salesman Problem: A Guided Tour of Combinatorial Optimization*, Wiley, New York, 1985.

LEVY, H., and LESSMAN, F. *Finite Difference Equations*, Macmillan, New York, 1961.

LEWIS, H. R., and PAPADIMITRIOU, C. H. *Elements of the Theory of Computation*, Prentice-Hall, Englewood Cliffs, N.J., 1981.

LICK, D. R. "A Sufficient Condition for Hamiltonian Connectedness," *J. Comb. Theory* 8 (1970), 444–445.

LIU, C. L. *Introduction to Combinatorial Mathematics*, McGraw-Hill, New York, 1968.

LIU, C. L. *Elements of Discrete Mathematics*, McGraw-Hill, New York, 1985.

MACMAHON, P. *Combinatory Analysis*, Vol. 1 (1915) and Vol. 2 (1916); reprinted in one volume by Chelsea, New York, 1960.

MARKOWSKY, G. "Best Huffman Trees," *Acta Inform.* 16 (1981), 363–370.

MAY, K. O. "The Origin of the Four Color Conjecture," *Isis* 56 (1965), 346–348.

MINIEKA, E. *Optimization Algorithms for Networks and Graphs*, Marcel Dekker, New York, 1978.

MOON, J. W. *Topics in Tournaments*, Holt, Rinehart and Winston, New York, 1968.

NEMHAUSER, G. L. "A Generalized Label Setting Algorithm for the Shortest Path between Specified Nodes," *J. Math. Anal. Appl.* 38 (1972), 328–334.

ORE, O. *Graphs and Their Uses*, New York, 1963. Random House, N.Y.

PAPADIMITRIOU, C. H. "The Complexity of Edge Traversing," *J. Assoc. Comput. Mach.* 23 (1976), 544–554.

PAPADIMITRIOU, C. H. and STEIGLITZ, K. *Combinatorial Optimization: Algorithms and Complexity*, Prentice-Hall, Englewood Cliffs, N.J., 1982.

POLYA, G. *Induction and Analogy in Mathematics*, Princeton Univ. Press, Princeton, N.J., 1963.

PRATT, V. "Every Prime Has a Succinct Certificate," *SIAM J. Comput.* 4 (1975), 214–220.

PRIM, R. C. "Shortest Connection Networks and Some Generalizations," *Bell System Tech. J.* 36 (1957), 1389–1401.

RALSTON, A. "De Bruijn Sequences—A Model Example of the Interaction of Discrete Mathematics and Computer Science," *Math. Mag.* 55 (1982), 131–143.

READ, R. C. "An Introduction to Chromatic Polynomials," *J. Combin. Theory* 4 (1968), 52–71.

REDEI, L. "Ein kombinatorischer Satz," *Acta Litt. Sci. Szegu* 7 (1934), 39–43.

REINGOLD, E. M., NIEVERGELT, J., and DEO, N. *Combinatorial Algorithms: Theory and Practice*, Prentice-Hall, Englewood Cliffs, N.J., 1977.

RIORDAN, J. *An Introduction to Combinatorial Analysis*, Princeton University Press, Princeton, N.J., 1958.

ROBBINS, H. E. "A Theorem on Graphs with an Application to a Problem of Traffic Control," *Amer. Math. Monthly* 46 (1939), 281–283.

ROBERTS, F. S. *Discrete Mathematical Models with Applications to Social, Biological and Environmental Problems*, Prentice-Hall, Englewood Cliffs, N.J., 1976.

ROBERTS, F. S. *Graph Theory and Its Applications to Problems of Society*, SIAM, Philadelphia, 1978.

ROBERTS, F. S. *Applied Combinatorics*, Prentice-Hall, Englewood Cliffs, N.J., 1984.

RONSE, C. *Feedback Shift-Registers*, Springer-Verlag, New York, 1982.

RYSER, H. J. *Combinatorial Mathematics*, Carus Mathematical Monographs No. 14, Mathematical Association of America, Washington D.C., 1963.

SCHRIJVER, A. *Theory of Linear and Integer Programming*, Wiley, New York, 1986.

SOMINSKIT, I. S. *The Method of Math Induction*, D. C. Heath, Boston, 1963.

STANAT, D. F., and McALLISTER, D. F. *Discrete Mathematics in Computer Science*, Prentice-Hall, Englewood Cliffs, N.J., 1977.

STANDISH, T. A. *Data Structure Techniques*, Addison-Wesley, Reading, Mass., 1980.

STOLL, R. R. *Set Theory and Logic*, W. H. Freeman, San Francisco, 1963.

SWAMY, M. N. S., and THULASIRAMAN, K. *Graphs, Networks and Algorithms*, Wiley, New York, 1981.

SYSLO, M. M., DEO, N., and KOWALIK, J. S. *Discrete Optimization Algorithms with Pascal Structures*, Prentice-Hall, Englewood Cliffs, N.J., 1983.

TABOURIER, Y. "All Shortest Distances in a Graph," *Discrete Math.* 4 (1973), 83–87.

TARJAN, R. E. "Depth First Search and Linear Graph Algorithms," *SIAM J. Comput.* 1 (1971), 146–160.

TOWNSEND, M. *Discrete Mathematics: Applied Combinatorics and Graph Theory*, Benjamin-Cummings, Menlo Park, Calif., 1987.

TUCKER, A. *Applied Combinatorics*, 2nd ed. Wiley, New York, 1984.

TYMOCZKO, T. "Computers, Proofs and Mathematicians: A Philosophical Investigation of the Four Color Proof," *Math. Mag.* 53 (1980), 131–138.

WARSHALL, S. "A Theorem on Boolean Matrices," *J. Assoc. Comput. Mach.* 9 (1962), 11–12.

WHITWORTH, W. A. *Choice and Chance* (reprint of the fifth edition, 1901), Hafner Press, New York, 1965.

WILF, H. S. *Algorithms and Complexity*, Prentice-Hall, Englewood Cliffs, N.J., 1986.

WILSON, R. J. *Introduction to Graph Theory*, 2nd ed. Longman Group, Harlow, Essex, England, 1979.

YEMELICHEV, V. A., KOVALEV, M. M., and KRAVTSOV, M. K. *Polytopes, Graphs and Optimisation*, Cambridge University Press, Cambridge, 1984.

Answers
to Selected Exercises

0.1. (a) $A \cup B = \{2, 3, 5, 6, 7, 9\}$ **(b)** $B \cap C = \{2, 6\}$ **(c)** $B - A = \{2, 6\}$
(d) $A - B = \{9\}$ **(e)** $C' = \{3, 5, 7, 9\}$ **(f)** X' empty set **(g)** Complement of the empty set X

0.3. $\{a, b, c, c\} = \{a, b, a, b, c\}$

0.5. (a) $A \times A = \{(3, 3), (4, 4), (3, 4), (4, 3)\}$ **(b)** $A \times B = \{(3, p), (3, q), (3, r), (4, p), (4, q), (4, r)\}$ **(c)** $B \times A = \{(p, 3), (p, 4), (q, 3), (q, 4), (r, 3), (r, 4)\}$ **(d)** $B \times B = \{(p, p), (q, q), (r, r), (p, q), (q, p), (p, r), (r, p), (q, r), (r, p)\}$

0.7. (a) $A \cup (B \times A) = \{3, 4, (p, 3), (p, 4), (q, 3), (q, 4), (r, 3), (r, 4)\}$
(b) $(A \times A) \cup (B \times B) = \{(3, 3), (4, 4), (3, 4), (4, 3), (p, 3), (p, 4), (q, 3), (q, 4), (r, 3), (r, 4)\}$

0.9. (a) $\{\{a\}\}$ **(b)** $\{\{a\}, \{b\}\}$ **(c)** $\{\{a\}, \{b\}, \{c\}\}, \{\{a, b\}, \{c\}\}, \{\{a, c\}, \{b\}\},$ and $\{\{b, c\}, \{a\}\}$

0.13. (a) Yes **(b)** Yes

0.19. One

0.23. At most eight

0.25. Both $(A - B)$ and $(B - A)$ are empty. This means that A is a subset of B and at the same time B is a subset of A. So $A = B$.

0.27. Yes. Let x be an arbitrary element of B. Case 1: Suppose x is not in A. Then x is in the symmetric difference of A and B. So x is in the symmetric difference of A and C. So x is in $A \cup C$ but not in $A \cap C$. So x is in C. Case 2: Suppose x is in A. Then x is not in the symmetric difference of A and B and therefore x is not in the symmetric difference of A and C. So x is in $A \cap C$ which implies x is in C. Thus in any case B is a subset of A. Likewise C is a subset of B.

0.29. **(a)** Both the domain and the codomain are R and the range is the set of all nonnegative real numbers. **(b)** No **(c)** No **(d)** $\{-2, 2\}$ **(e)** The union of the two closed intervals I and J where $I = \{x : -2 \le x \le -1\}$ and $J = \{x : 1 \le x \le 2\}$

0.31. The function f is not a surjection. The inverse function $g(n) = (n - 5)/2$ where n is in $f(N)$ is a surjection.

0.33. **(a)** 3.3.3.3 **(b)** 0 **(c)** (3.3.3.3) − (3.2.2.2.2) + 3 **(d)** 1.1.3.3

0.35. **(a)** Domain = the set of all integers, range = $\{0, 1, 2, 3, 4, 5, 6, 7, 8, 9\}$ **(b)** Domain = the set of all integers, range = the set of all positive integers

0.37. $\{(p, 1), (q, 1), (r, 2)\}$

0.39. $\{(p, p), (q, q), (r, p)\}$ when n is even and $\{(p, q), (q, p), (r, q)\}$ when n is odd

0.43. The four constants satisfy the equation $ad + b = bc + d$.

0.49. **(a)** 2 **(b)** 3

0.51. Draw five small circles on the left side one below the other such that no two circles touch each other and label them 1, 2, 3, 4, and 5. Then draw four small circles one below the other on the right side and label them a, b, c, and d. Draw arrows (1) from 1 to a, (2) from 1 to b, (3) from 3 to c, (4) from 4 to d, (5) from 5 to d, and (6) from 5 to c.

0.53. No in all cases except in (c)

0.55. $R^2 = \{(a, a), (a, b), (b, a), (b, b), (c, a), (c, b)\}$
$R^3 = \{(a, a), (a, b), (b, a), (b, b), (c, a), (c, b)\}$

0.62. Reflexive, not symmetric, antisymmetric, transitive

0.64. **(a)** This is an equivalence relation. The corresponding partition is $\{\{1, 3\}, \{2\}, \{4\}\}$. **(b)** This is not an equivalence relation. **(c)** This is an equivalence relation with the partition $\{\{1, 2\}, \{2\}, \{3\}, \{4\}\}$.

0.68. **(a)** $\{5k : k \in Z\}$ **(b)** $\{1 + 5k : k \in Z\}$ **(c)** $\{2 + 5k : k \in Z\}$

0.70. The preimages of the elements of the range of f

0.78. The set S has 27 pairs in all. There are 8 pairs in which the first element is the empty set and the second element is any subset of X. There are 12 pairs in which the first element is a singleton set and 6 pairs in which the first element is a set of two elements. Finally, $(\{a, b, c\}, \{a, b, c\})$ is in S.

0.80. $\{1, 2, 4, 8\}$ and $\{1, 3, 6\}$ are two chains.

0.82. **(a)** 2, 3 **(b)** 16, 24 **(c)** 12, 24

0.87. The induction hypothesis $P(n)$ is the statement that the sum $1/1 \cdot 2 + 1/2 \cdot 3 + 1/3 \cdot 4 + \cdots + 1/n(n + 1)$ equals $n/(n + 1)$. The aim is to prove that this statement is true for all n. Obviously, $P(1)$ is true since $1/1 \cdot 2 = 1/(1 + 1)$. So the basis step is proved. Next we have to prove the induction step: if $P(k)$ is true for any k, then $P(k + 1)$ is true as well. It is easy to see that $P(k + 1)$ is true because $k/(k + 1) + 1/(k + 1)(k + 2)$ is equal to $(k + 1)/(k + 2)$.

0.105. $f(1) = 1$ and $f(n) = n + f(n - 1)$

0.107. It is a tautology.

0.109. Neither

0.111.

p'	q'	$(p' \vee q')$	q	$(p' \vee q)' \leftrightarrow q$
T	T	T	F	F
T	F	T	T	T
F	T	T	F	F
F	F	F	T	F

0.113. **(a)** Satisfiable in all cases except when p, q, and r are true and s and t are false   **(b)** Satisfiable in all cases except when p and q are true and r is false

Chapter 1

1.1. 10^9

1.3. **(a)** 16   **(b)** 8   **(c)** 102   **(d)** 28

1.5. **(a)** 2^8 **(b)** 16   **(c)** 48   **(d)** 112

1.7. There are $26 \cdot 25 \cdot 24 \cdot 23 \cdot 22$ ways.

1.9. $n(n - 1)$

1.11. $26 + (26)(36) + (26)(36)^2 + (26)(36)^3 + (26)(36)^4 + (26)(36)^5$

1.13. **(a)** 12   **(b)** 144   **(c)** 72

1.17. $(6!) = 720$; then $(7!) = (7)(720) = 5040$ and $(8!) = (8)(5040) = 40{,}320$

1.19. $n = 23$

1.21. **(a)** $(4!)(5!)(6!)$ **(b)** $(3!)(4!)(5!)(6!)$

1.23. 5040

1.25. 86,400

1.27. **(a)** $(10!)/(4!)(4!)(2!)$ **(b)** $(12!)/(5!)(4!)(3!)$

1.31. **(a)** $P(11, 9)$ **(b)** $P(11; 2, 3, 4)$

1.33. **(a)** 512   **(b)** 84   **(c)** 36

1.35. There are 504 ways.

1.37. $C(n, r) \cdot (r - 1)!$

1.39. The number of ways is r where $r = C(14, 8) \cdot (7!) \cdot (6!)$.

1.41. $P(7; 4, 2, 1) \cdot C(8, 4)$

1.47. $P(10; 2, 3, 4, 1) \cdot C(23, 1)$

1.49. **(a)** $C(10, 6) \cdot C(12, 6)$ **(b)** $C(12, 7) \cdot C(10, 5) + C(12, 8) \cdot C(10, 4) + C(12, 9) \cdot C(10, 3) + C(12, 10) \cdot C(10, 2) + C(12, 11) \cdot C(10, 1) + C(12, 12) \cdot C(10, 0)$

1.57. **(a)** $(18!)/[(4!) \cdot (1!) \cdot (4!)^4 \cdot (2!)^1]$ **(b)** $(18!)/[(2) \cdot (2) \cdot (5!)^2 \cdot (4!) \cdot (2!)^2]$ **(c)** $C(18; 7, 6, 5)$

1.58. **(a)** 840   **(b)** 7^4 **(c)** 7^9

1.61. $C(28, 4)$

1.63. **(a)** $C(14, 4)$ **(b)** $C(9, 4)$ **(c)** $C(8, 3) + C(6, 3) + C(4, 3)$

1.65. $C(10, 4)$

1.69. $C((r - p) + n - 1, n - 1)$, where $p = p_1 + p_2 + \cdots + p_n$

1.72. $C(15, 5) - C(8, 5)$

1.74. (a) 25 (b) 27

1.78. 275

1.80. (a) Let $X = \{1, 2, \ldots, n\}$ and A be the set of numbers in X that are not squarefree. The number of squarefree integers in X is obviously $n - N(A)$. To compute $N(A)$, proceed as follows. Let $P = \{p_1, p_2, \ldots, p_r\}$, where each p_i is a prime number that does not exceed the square root of n. Let A_i be the set of those numbers in X that are divisible by the square of p_i. Compute S_i $(i = 1, 2, \ldots, r)$ as in Theorem 1.6.1. Then $N(A)$ is equal $S_1 - S_2 + \cdots + (-1)^{r-1}$. (b) $100 - (42 - 3 + 0 - 0)$

1.82. (a) Number of permutations $= 6! = 720$, number of derangements $= D_6 = 265$ (b) $265/720$ (c) $[C(6, 1) \cdot D_5]/(6!) = 0.366667$ (d) $1 - 0.366667$ (e) $[C(6, 2) \cdot D_4]/(6!)$ (f) $1/720$

1.84. The answer is 0.

1.86. There are 120 ways.

Chapter 2

2.1. (a) $1 + x + x^2 + x^3$ (b) $x^4 + x^5 + x^6 + \cdots$ (c) $1 + x + x^2 + x^3 + \cdots$ (d) $1 - x + x^2 - x^3 + x^4 - \cdots$

2.3. (a) $\{16, 32, 24, 8, 1, 0, 0, 0, \ldots\}$ (b) $\{1, 1, 3/2, 1/3!, 1/4!, 1/5!, \ldots\}$ (c) $\{0, 0, 0, 1, 1, 1, 1, \ldots\}$

2.5. $C(15, 8)$

2.7. $(x + x^2 + x^3 + \cdots + x^r)^4$

2.9. $C(18, 3) - 4 \cdot C(12, 3) + 6 \cdot C(6, 3)$

2.11. $C(12, 2) - 3 \cdot C(6, 2)$

2.13. There are 18 ways.

2.17. Coefficient of the tenth power of x in $f(x)$, where $f(x) = (x + x^2)(x + x^2 + x^3)(x + x^2 + \cdots)^2$

2.19. The function is $x^6 (1 - x^4)^3 \cdot (1 - x)^{-4}$.

2.21. 10

2.23. 6

2.25. 30

2.27. $(x^4 + x^8 + \cdots)(1 + x^3 + x^6 + \cdots)(1 + x^2 + x^4 + \cdots)(1 + x + x^2 + \cdots)$

2.29. $C(n - r + 1, r)$

2.37. The number of such numbers is t, where $t = (1/4)(2^r + 2^r)$ where r is even. If r is odd, t is necessarily zero.

2.41. $[(e^x + e^{-x})/2 - 1](e^x - 1)^4$

Chapter 3

3.1. $f(n) = f(n - 1) + n$ where $f(1) = 2$; $f(9) = 46$

3.3. $f(n) = 2f(n - 1) + 1$ with $f(1) = 1$; $f(n) = 2^n - 1$

3.5. $f(n) = f(n - 1) + f(n - 2)$ with $f(1) = 2$ and $f(2) = 3$

3.8. $f(n) = 2f(n - 1)$ with $f(1) = 2$

3.10. (a) $k = 2$ (b) The initial conditions are not consecutive.

3.12. $f(n) = 1 + n + 2^n$

3.14. The characteristic polynomial is $(x - 1)(x - 2)^2(x - 3)$ and $f(n) = A + B \cdot 2^n + C \cdot n \cdot 2^n + D \cdot 3^n$ is the general solution of the relation $f(n + 4) = 8f(n + 3) - 23f(n + 2) + 28f(n + 1) - 12f(n)$.

3.16. $g(n) = A(-1)^n + B(m - 1)^n$, where $A = (-1)/m$ and $B = 1/m$ and $f(n) = (m - 1)g(n - 1)$

3.18. $f(n) = A + B \cdot (3)^n - 8n$, where $A = 1$ and $B = 3$

3.20. $A(4)^n + 5n(4)^n$

3.22. $A(2)^n + B \cdot n \cdot (2)^n + (1/2) \cdot n^2 \cdot (2)^n$

3.24. $p = -5$, $q = 6$, and $r = 8$

3.26. $f(n) =$ coefficient of x^n in $g(x) = 2/(1 + x) - 1/(1 - x)^2 + 2/(1 - x)^3$

3.28. $f(n) = 5n^2 - 4n$

3.30. $f(n) = d + c \log n$

3.32. The relation is $g(n) = 7g(n/2) + 18(n^2)^2$ with $g(1) = 0$. The solution is $6 \cdot n^r - 6 \cdot n^2$, where $r = \log 7$.

3.34. **(a)** $f(n) = 2f(n - 1)$ with $f(1) = 0$ **(b)** $f(n) = f(n/2) + (n - 1)$ with $f(1) = 0$ **(c)** The two solutions are (1) $n(n - 1)/2$ and (2) $2n - \log n - 2$. When $n > 3$, the second is more efficient than the first.

Chapter 4

4.1. $W = \{1, 3\}$

4.3. (b) It is possible to draw K_4 such that no edges intersect. It is not possible to do so for K_5.

4.5. Draw a simple graph as suggested. There are edges between every pair of cities except between Boston and Moscow, and between Boston and Prague.

4.7. (a) Two **(b)** Two

4.9. pq

4.11. Suppose that the arcs of the digraph are $(1, 2)$, $(1, 3)$, $(1, 5)$, $(2, 3)$, $(3, 4)$, $(3, 5)$, and $(4, 5)$.

(a) The adjacency matrix $A = (a_{ij})$ is a 5×5 matrix in which $a_{12} = a_{13} = a_{15} = a_{23} = a_{34} = a_{35} = a_{45} = 1$ and all the other elements are zero.

(b) (Indegree of vertex 1) $= 0$, (outdegree of vertex 1) $= 3$
    (Indegree of vertex 2) $= 1$, (outdegree of vertex 2) $= 1$
    (Indegree of vertex 3) $= 2$, (outdegree of vertex 3) $= 2$
    (Indegree of vertex 4) $= 1$, (outdegree of vertex 4) $= 1$
    (Indegree of vertex 5) $= 3$, (outdegree of vertex 5) $= 0$

(c) $(1, 2)$ $(1, 3)$ $(1, 5)$ $(2, 3)$ $(3, 4)$ $(3, 5)$ $(4, 5)$

$$
\begin{array}{c}
1 \\
2 \\
3 \\
4 \\
5
\end{array}
\begin{bmatrix}
-1 & -1 & -1 & 0 & 0 & 0 & 0 \\
1 & 0 & 0 & -1 & 0 & 0 & 0 \\
0 & 1 & 0 & 1 & -1 & -1 & 0 \\
0 & 0 & 0 & 0 & 1 & 0 & -1 \\
0 & 0 & 1 & 0 & 0 & 1 & 1
\end{bmatrix}
$$

4.13. The graph looks like a bracelet or ring studded with stones such that each vertex is represented by a stone. Such a graph with n vertices is denoted by Z_n and is called an "odd hole" if n is odd and more than 3. It is an "even hole" if n is even and more than 3.

4.15. (a) $G = (V, E)$, where $V = \{1, 2, 3, 4\}$ and $E = \{\{1, 2\}, \{3, 4\}\}$
(b) $G = (V, E)$, where $V = \{1, 2, 3, 4\}$ and $E = \{\{1, 2\}, \{2, 3\}, \{3, 4\}, \{4, 1\}\}$
(c) $(nr)/2$, so at least one of the two numbers is even.
4.19. (a) $1 \to 2 \to 3 \to 4 \to 5 \to 2 \to 6$ **(b)** $1 \to 2 \to 3 \to 4 \to 5 \to 6$
(c) $2 \to 3 \to 4 \to 5 \to 2$ **(d)** One **(e)** $\{1\}, \{6\}, \{2, 3, 4, 5\}$ **(f)** This is a 6×6 matrix in which (1) all the elements in row 6 are 0, (2) all the elements in column 1 are 0 except the first element, and (3) all the remaining elements are 1.
4.21. $G = (V, E)$, where $V = \{1, 2, 3, 4, 5\}$ and $E = \{\{1, 2\}, \{2, 3\}, \{3, 4\}, \{4, 2\}, \{2, 5\}, \{5, 1\}\}$
4.23. If the graph is connected, every element is 1. More generally, if G has n vertices and k components, the $n \times n$ reachability matrix A of G will have k submatrices along the diagonal of A such that every element in each submatrix is 1 and every other element in A is zero. If G_i is a component of G with n_i vertices, the submatrix corresponding to this component will be a $n_i \times n_i$ matrix.
4.27. G is not connected.
4.29. The three vertices at the top are marked 1, 2, and 3. The three vertices are marked 8, 9, and 4 from the left to the right. The three vertices at the bottom are marked 7, 6, and 5 from the left to the right. The arcs (i, j) in which $i < j$ are $(1, 2)$, $(2, 3)$, $(3, 4)$, $(4, 5)$, $(5, 6)$, $(6, 7)$, $(7, 8)$, and $(8, 9)$. The remaining arcs are (i, j) where $i > j$.

Chapter 5

5.3. There will be a directed path from every vertex to every vertex.
5.5. No
5.7. The word starts with C since its row sum 2 equals its column sum plus 1. The word ends with B since its column sum equals its row sum plus 1. For the other two letters row sum equals column sum. The row sums of A and D are 2 and 3. Thus the frequencies of A, B, C, and D are 2, 2, 2, and 3, respectively. Draw a digraph with 4 vertices A, B, C, D. Draw an arc from a letter X (X is one of these four letters) to a letter Y (Y is also one of these four letters, the letters X and Y need not be distinct) if the element in the matrix corresponding to row X and column Y is 1. In the resulting digraph there will be a directed Eulerian path (not necessarily unique) from C to B. Any such path will give a word.
5.9. (a) The digraph is $G = (V, A)$ where $V = \{0, 1, 2\}$ and A is the Cartesian product $V \times V$. An arc from vertex i to vertex j is assigned the word ij. The consecutive arcs in the following sequence will give an Eulerian circuit starting at vertex 0 and ending in vertex 0: $< 00 \quad 01 \quad 11 \quad 12 \quad 22 \quad 21 \quad 10 \quad 02 \quad 21 >$. The first letters of these nine words define the the de Bruijn sequence $B(3, 2) =$ $< a_1 \quad a_2 \quad a_3 \quad a_4 \quad a_5 \quad a_6 \quad a_7 \quad a_8 \quad a_9 > = < 0 \quad 0 \quad 1 \quad 1 \quad 2 \quad 2 \quad 1 \quad 0 \quad 2 >$ and any two-letter word using the three symbols 0, 1, and 2 is of the form $a_i a_{i+1}$ where i is any integer such that $1 \le i \le 9$ and the addition of the subscripts is modulo 9.
5.11. $V = \{1, 2, 3, 4\}$; $E = \{\{1, 2\}, \{2, 3\}, \{3, 4\}, \{4, 1\}, \{1, 3\}\}$
5.13. $V = \{1, 2, 3, 4\}$; $E = \{\{1, 2\}, \{1, 3\}, \{1, 4\}\}$
5.15. The number of edges in a Hamiltonian cycle of a graph with n vertices is n.

The number of edges in any cycle of a bipartite graph is even. So a bipartite graph with an odd number of vertices cannot be Hamiltonian.

5.17. Yes, by definition. The converse is not true, for consider the counterexample with $V = \{1, 2, 3\}$ and $A = \{(1, 2), (2, 3)\}$.

Chapter 6

6.1. $m = n - k$

6.3. 19

6.5. 24

6.7. It is a bridge.

6.9. Not necessary. If the vertex set of the complete graph with five vertices is $\{1, 2, 3, 4, 5\}$, then T and T' are two distinct spanning trees with $E = \{\{1, 2\}, \{2, 3\}, \{3, 4\}, \{4, 5\}\}$ and $E' = \{\{1, 3\}, \{3, 5\}, \{2, 5\}, \{2, 4\}\}$.

6.13. The tree is $T = (V, E)$, where $V = \{1, 2, 3, 4, 5, 6, 7, 8, 9\}$ and $E = \{(1, 8), (2, 8), (3, 7), (4, 7), (5, 7), (7, 6), (8, 6), (6, 9)\}$.

6.15. ABRACADABRA

6.17. $A = 110, B = 00, C = 1110, D = 1111, E = 10, R = 01$. The word is

110 00 01 110 1110 110 1111 110 00 01 110

The length is at most 31.

6.19. **(a)** $n + 1 = 14$ and $3 < \log 14 < 4$, so $m = 3$. So the height cannot be less than 3. **(b)** The floor of 13/2 is 6, so the height is not more than 6.

6.21. The tree is rooted at H. The left tree has G, B, A, and C. The right tree has P, R, and Y. The left child of G is B. There is no right child for G. The left child of B is A and the right child of B is C. On the right subtree rooted at H, no vertex has a left child.

Chapter 7

7.5. The edges are $\{1, 5\}, \{1, 4\}, \{3, 4\}, \{2, 6\}$, and $\{5, 6\}$. The weight of the tree is 31.

Chapter 8

8.1.

$$A^{(7)} = \begin{bmatrix} 0 & 3 & 2 & 2 & 3 & 4 & 2 \\ - & 0 & -1 & -1 & 2 & 1 & 0 \\ - & 1 & 0 & 0 & 3 & 2 & 1 \\ - & 1 & 0 & 0 & 3 & 2 & 1 \\ - & 0 & -1 & -1 & 0 & 1 & -1 \\ - & 1 & 0 & 0 & 3 & 2 & 0 \end{bmatrix} \quad P^{(7)} = \begin{bmatrix} 1 & 5 & 5 & 5 & 5 & 5 & 5 \\ 1 & 2 & 3 & 4 & 5 & 3 & 7 \\ 1 & 6 & 3 & 6 & 5 & 6 & 7 \\ 1 & 3 & 3 & 4 & 3 & 3 & 3 \\ 1 & 2 & 2 & 2 & 2 & 6 & 2 \\ 1 & 6 & 6 & 6 & 6 & 6 & 7 \end{bmatrix}$$

8.3. The arcs of this tree rooted at vertex 1 are $(1, 5), (5, 7), (7, 6), (5, 2), (2, 3)$, and $(3, 4)$.

8.5. In $A^{(4)}$ the element corresponding to the fourth row and second column is 4. So the S.D. from 4 to 2 without touching 5, 6, or 7 will be 4. Furthermore, in $p^{(4)}$, the element corresponding to the fourth row and second column is 2, indicating that we go straight from 4 to 2.

8.7. Replace the tractor at the end of the first year. The total cost will be $12 + 32 = 44$.

Index